mathematics and found total happiness

SURREAL
NUMBERS

ADDISON-WESLEY PUBLISHING COMPANY

London · Amsterdam · Don Mills, Ontario · Sydney

This book has been set in Modern Extended #7 with Albertus chapter headings. The cover and illustrations were designed by Jill C. Knuth.

ISBN 0-201-03812-9
CDEFGHIJ-AL-79876

CONTENTS

1

THE ROCK

A. Bill, do you think you've found yourself?

B. What?

A. I mean—here we are on the edge of the Indian Ocean, miles away from civilization. It's been months since we ran off to avoid getting swept up in the system, and "to find ourselves." I'm just wondering if you think we've done it.

B. Actually, Alice, I've been thinking about the same thing. These past months together have been really great—we're completely free, we know each other, and we feel like real people again instead of like machines. But lately I'm afraid I've been missing some of the things we've "escaped" from. You know, I've got this fantastic craving for a book to read—*any* book, even a textbook, even a math textbook. It sounds crazy, but I've been lying here wishing I had a crossword puzzle to work on.

A. Oh, c'mon, not a crossword puzzle; that's what your *parents* like to do. But I know what you mean, we need some mental stimulation. It's kinda like the end of summer vacations when we were kids. In May every year we couldn't wait to get out of school, and the days simply dragged on until vacation started, but by September we were real glad to be back in the classroom.

B. Of course, with a loaf of bread, a jug of wine, and thou beside me, these days aren't exactly "dragging on." But I think maybe the most important thing I've learned on this trip is that the simple, romantic life isn't enough for me. I need something complicated to think about.

A. Well, I'm sorry I'm not complicated enough for you. Why don't we get up and explore some more of the beach? Maybe we'll find some pebbles or something that we can use to make up some kind of a game.

B. (sitting up) Yeah, that's a good idea. But first I think I'll take a little swim.

A. (running toward the water) Me, too—bet you can't catch me!

. . . .

4

B. Hey, what's that big black rock half-buried in the sand over there?

A. Search me, I've never seen anything like it before. Look, it's got some kind of graffiti on the back.

B. Let's see. Can you help me dig it out? It looks like a museum piece. Unnh! Heavy, too. The carving might be some old Arabian script...no, wait, I think it's maybe Hebrew; let's turn it around this way.

A. Hebrew! Are you sure?

B. Well, I learned a lot of Hebrew when I was younger, and I can almost read this....

A. I heard there hasn't been much archaeological digging around these parts. Maybe we've found another Rosetta Stone or something. What does it say, can you make anything out?

B. Wait a minute, gimme a chance.... Up here at the top right is where it starts, something like "In the beginning everything was void, and...."

A. Far out! That sounds like the first book of Moses, in the Bible. Wasn't he wandering around Arabia for forty years with his followers before going up to Israel? You don't suppose....

B. No, no, it goes on much different from the traditional account. Let's lug this thing back to our camp, I think I can work out a translation.

A. Bill, this is wild, just what you needed!

B. Yeah, I did say I was dying for something to read, didn't I. Although this wasn't exactly what I had in mind! I can hardly wait to get a good look at it—some of the things are kinda strange, and I can't figure out whether it's a story or what. There's something about numbers, and....

A. It seems to be broken off at the bottom; the stone was originally longer.

B. A good thing, or we'd never be able to carry it. Of course it'll be just our luck to find out that the message is getting interesting, right when we come to the broken place.

A. Here we are. I'll go pick some dates and fruit for supper while you work out the translation. Too bad languages aren't my thing, or I'd try to help you.

. . . .

B. Okay, Alice, I've *got* it. There are a few doubtful places, a couple signs I don't recognize; you know, maybe some obsolete word forms. Overall I think I know what it says, though I don't know what it means. Here's a fairly literal translation:

> In the beginning, everything was void, and J. H. W. H. Conway began to create numbers. Conway said, "Let there be two rules which bring forth all numbers large and small. This shall be the first rule: Every number corresponds to two sets of previously created numbers, such that no member of the left set is greater than or equal to any member of the right set. And the second rule shall be this: One number is less than or equal to another number if and only if no member of the first number's left set is greater than or equal to the second number, and no member of the second number's right set is less than or equal to the first number." And Conway examined these two rules he had made, and behold! they were very good.
>
> And the first number was created from the void left set and the void right set. Conway called this number "zero,"

and said that it shall be a sign to separate positive numbers from negative numbers. Conway proved that zero was less than or equal to zero, and he saw that it was good. And the evening and the morning were the day of zero. On the next day, two more numbers were created, one with zero as its left set and one with zero as its right set. And Conway called the former number "one," and the latter he called "minus one." And he proved that minus one is less than but not equal to zero and zero is less than but not equal to one. And the evening...

That's where it breaks off.

A. Are you *sure* it reads like that?

B. More or less. I dressed it up a bit.

A. But "Conway"...that's not a Hebrew name. You've got to be kidding.

B. No, honest. Of course the old Hebrew writing doesn't show any vowels, so the real name might be Keenawu or something; maybe related to the Khans? I guess not. Since I'm translating into English, I just used an English name. Look, here are the places where it shows up on the stone. The J. H. W. H. might also stand for "Jehovah."

A. No vowels, eh? So it's real... But what do you think it means?

B. Your guess is as good as mine. These two crazy rules for numbers. Maybe it's some ancient method of arithmetic that's been obsolete since the wheel was invented. It might be fun to figure them out, tomorrow; but the sun's going down pretty soon so we'd better eat and turn in.

A. Okay, but read it to me once more. I want to think it over, and the first time I didn't believe you were serious.

B. (pointing) "In the beginning,...."

A. I think your Conway Stone makes sense after all, Bill. I was thinking about it during the night.

B. So was I, but I dozed off before getting anywhere. What's the secret?

A. It's not so hard, really; the trouble is that it's all expressed in words. The same thing can be expressed in symbols and then you can see what's happening.

B. You mean we're actually going to use the New Math to decipher this old stone tablet.

A. I hate to admit it, but that's what it looks like. Here, the first rule says that every number x is really a pair of sets, called the left set x_L and the right set x_R:

$$x = (x_L, x_R).$$

B. Wait a sec, you don't have to draw in the sand, I think we still have a pencil and some paper in my backpack. Just a minute...Here, use this.

A. $x = (x_L, x_R).$

These x_L and x_R are not just numbers, they're *sets* of numbers; and each number in the set is itself a pair of sets, and so on.

B. Hold it, your notation mixes me up. I don't know what's a set and what's a number.

A. Okay, I'll use capital letters for sets of numbers and small letters for numbers. Conway's first rule is that

$$x = (X_L, X_R), \qquad \text{where} \qquad X_L \not\geq X_R. \tag{1}$$

This means if x_L is any number in X_L and if x_R is any number in X_R, they must satisfy $x_L \not\geq x_R$. And that means x_L is not greater than or equal to x_R.

B. (scratching his head) I'm afraid you're still going too fast for me. Remember, you've already got this thing psyched out, but I'm still at the beginning. If a number is a pair of sets of numbers, each of which is a pair of sets of numbers, and so on and so on, how does the whole thing get started in the first place ?

10

A. Good point, but that's the whole beauty of Conway's scheme. Each element of X_L and X_R must have been created previously, but on the first day of creation there weren't any previous numbers to work with; so both X_L and X_R are taken as the empty set!

B. I never thought I'd live to see the day when the empty set was meaningful. That's really creating something out of nothing, eh? But is $X_L \not\geq X_R$ when X_L and X_R are both equal to the empty set? How can you have something unequal itself?

Oh yeah, yeah, that's okay since it means no *element* of the empty set is greater than or equal to any element of the empty set—it's a true statement because there *aren't* any elements in the empty set.

A. So everything gets started all right, and that's the number called zero. Using the symbol $\emptyset$ to stand for the empty set, we can write

$$0 = (\emptyset, \emptyset).$$

B. Incredible.

A. Now on the second day, it's possible to use 0 in the left or right sets, so Conway gets two more numbers

$$-1 = (\emptyset, \{0\}) \qquad \text{and} \qquad 1 = (\{0\}, \emptyset).$$

B. Let me see, does this check out? For -1 to be a number, it has to be true that no element of the empty set is greater than or equal to 0. And for 1, it must be that 0 is not greater than any element of the empty set. Man, that empty set sure gets around! Someday I think I'll write a book called *Properties of the Empty Set*.

A. You'd never finish.

> If X_L or X_R is empty, the condition $X_L \not\geq X_R$ is true no matter *what* is in the other set. This means that infinitely many numbers are going to be created.

B. Okay, but what about Conway's second rule?

A. That's what you use to tell whether $X_L \not\geq X_R$, when both sets are nonempty; it's the rule defining less-than-or-equal. Symbolically,

$$x \leq y \quad \text{means} \quad X_L \not\geq y \quad \text{and} \quad x \not\geq Y_R. \quad (2)$$

B. Wait a minute, you're way ahead of me again. Look, X_L is a set of numbers, and y is a number, which means a pair of sets of numbers. What do you mean when you write $X_L \not\geq y$?

A. I mean that every element of X_L satisfies $x_L \not\geq y$. In other words, no element of X_L is greater than or equal to y.

B. Oh, I see, and your rule (2) says also that x is not greater than or equal to any element of Y_R. Let me check that with the text...

A. The Stone's version is a little different, but $x \leq y$ must mean the same thing as $y \geq x$.

B. Yeah, you're right. Hey, wait a sec, look here at these carvings off to the side:

$$\bullet = \langle : \rangle$$

$$\mathbf{|} = \langle \bullet : \rangle$$

$$\rule{12pt}{2pt} = \langle : \bullet \rangle$$

12

These are the symbols I couldn't decipher yesterday, and your notation makes it all crystal clear! Those double dots separate the left set from the right set. You must be on the right track.

A. Wow, equal signs and everything! That stone-age carver must have used ▬ to stand for -1; I almost like his notation better than mine.

B. I bet we've underestimated primitive people. They must have had complex lives and a need for mental gymnastics, just like us—at least when they didn't have to fight for food and shelter. We always oversimplify history when we look back.

A. Yes, but otherwise how could we look back?

B. I see your point.

A. Now comes the part of the text I don't understand. On the first day of creation, Conway "proves" that $0 \leq 0$. Why should he bother to prove that something is less than or equal to itself, since it's obviously equal to itself. And then on the second day he "proves" that -1 is not equal to 0; isn't that obvious without proof, since -1 is a different number?

B. Hmm. I don't know about you, but I'm ready for another swim.

A. Good idea. That surf looks good, and I'm not used to so much concentration. Let's go!

3

B. An idea hit me while we were paddling around out there. Maybe my translation *isn't* correct.

A. What? It *must* be okay, we've already checked so much of it out.

B. I know; but now that I think of it, I wasn't quite sure of that word I translated "equal to." Maybe it has a weaker meaning, "similar to" or "like." Then Conway's second rule becomes "One number is less than or *like* another number if and only if..." And later on, he proves that zero is less than or *like* zero, minus one is less than but not like zero, and so forth.

A. Oh, right, that must be it, he's using the word in an abstract technical sense that must be defined by the rules. So of *course* he wants to prove that 0 is less than or like 0, in order to see that his definition makes a number "like" itself.

B. So does his proof go through? By rule (2), he must show that no element of the empty set is greater than or like 0, and that 0 is not greater than or like any element of the empty set...Okay, it works, the empty set triumphs again.

A. More interesting is how he could prove that -1 is *not* like 0. The only way I can think of is that he proved that 0 is not less-than-or-like -1. I mean, we have rule (2) to tell whether one number is less than or like another; and if x is not less-than-or-like y, it isn't less than y and it isn't like y.

B. I see, we want to show that $0 \leq -1$ is false. This is rule (2) with $x = 0$ and $Y_R = \{0\}$, so $0 \leq -1$ if and only if $0 \not\geq 0$. But 0 *is* ≥ 0, we know that, so $0 \not\leq -1$. He was right.

A. I wonder if Conway also tested -1 against 1; I suppose he did, although the rock doesn't say anything about it. If the rules are any good, it should be possible to prove that -1 is less than 1.

B. Well, let's see: -1 is $(\emptyset, \{0\})$ and 1 is $(\{0\}, \emptyset)$, so once again the empty set makes $-1 \leq 1$ by rule (2). On the other hand, $1 \leq -1$ is the same as saying that $0 \not\geq -1$ and $1 \not\geq 0$, according to rule (2), but we know that both of these are false. Therefore $1 \not\leq -1$, and it must be that $-1 < 1$. Conway's rules seem to be working.

16

A. Yes, but so far we've been using the empty set in almost every argument, so the full implications of the rules aren't clear yet. Have you noticed that almost everything we've proved so far can be put into a framework like this: "If X and Y are any sets of numbers, then $x = (\emptyset, X)$ and $y = (Y, \emptyset)$ are numbers, and $x \leq y$."

B. It's neat the way you've just proved infinitely many things, by looking at the pattern I used in only a couple of cases. I guess that's what they call abstraction, or generalization, or something. But can you also prove that your x is strictly *less* than y? This was true in all the simple cases and I bet it's true in general.

A. Uh huh...Well no, not when X and Y are both empty, since that would mean $0 \not\leq 0$. But otherwise it looks very interesting. Let's look at the case when X is the empty set, and Y is not empty; is it true that 0 is less than $(Y, \emptyset)$?

B. If so, then I'd call $(Y, \emptyset)$ a "positive" number. That must be what Conway meant by zero separating the positive and negative numbers.

A. Yes, but look. According to rule (2), we will have $(Y, \emptyset) \leq 0$ if and only if no member of Y is greater than or like 0. So if, for example, Y is the set $\{-1\}$, then $(Y, \emptyset) \leq 0$. Do you want positive numbers to be ≤ 0?

Too bad I didn't take you up on that bet.

B. Hmm. You mean $(Y, \emptyset)$ is going to be positive only when Y contains some number that is zero or more. I suppose you're right. But at least we now understand everything that's on the stone.

A. Everything up to where it's broken off.

B. You mean...?

A. I wonder what happened on the *third* day.

B. Yes, we should be able to figure that out, now that we know the rules. It might be fun to work out the third day, after lunch.

A. You'd better go catch some fish; our supply of dried meat is getting kinda low. I'll go try and find some coconuts.

4 BAD NUMBERS

B. I've been working on that Third Day problem, and I'm afraid it's going to be pretty hard. When more and more numbers have been created, the number of possible sets goes up fast. I bet that by the seventh day, Conway was ready for a rest.

A. Right. I've been working on it too and I get seventeen numbers on the third day.

B. Really? I found nineteen; you must have missed two. Here's my list:

$$\langle : \rangle \ \langle - : \rangle \ \langle \bullet : \rangle \ \langle \mathbf{I} : \rangle \ \langle - \bullet : \rangle \ \langle - \mathbf{I} : \rangle \ \langle \bullet \mathbf{I} : \rangle$$

$$\langle - \bullet \mathbf{I} : \rangle \ \langle : - \rangle \ \langle : \bullet \rangle \ \langle : \mathbf{I} \rangle \ \langle : - \bullet \rangle \ \langle : - \mathbf{I} \rangle$$

$$\langle : \bullet \mathbf{I} \rangle \ \langle : - \bullet \mathbf{I} \rangle \ \langle - : \bullet \rangle \ \langle \bullet : \mathbf{I} \rangle \ \langle - \bullet : \mathbf{I} \rangle \ \langle - : \bullet \mathbf{I}$$

A. I see you're using the Stone's notation. But why did you include $\langle : \rangle$? That was created already on the first day.

B. Well, we have to test the new numbers against the old, in order to see how they fit in.

A. But I only considered *new* numbers in my list of seventeen, so there must actually be *twenty* different at the end of the third day. Look, you forgot to include

$$\langle - : \mathbf{I} \rangle$$

in your list.

B. (blinking) So I did. Hmm . . . 20 by 20, that's 400 different cases we'll have to consider in rule (2). A lot of work, and not much fun either. But there's nothing else to do, and I know it'll bug me until I know the answer.

A. Maybe we'll think of some way to simplify the job once we get started.

B. Yeah, that would be nice . . .

Well, I've got one result, 1 is less than $(\{1\}, \emptyset)$. First I had to prove that $0 \not\geq (\{1\}, \emptyset)$.

A. I've been trying out a different approach. Rule (2) says we have to test every element of X_L to see that it isn't greater

22

than or like y, but it shouldn't be necessary to make so many tests. If any element of X_L is $\geq y$, then the *largest* element of X_L ought to be $\geq y$. Similarly, we need only test x against the *smallest* element of Y_R.

B. Yeah, that oughta be right . . . I can prove that 1 is less than $(\{0, 1\}, \emptyset)$ just like I proved it was less than $(\{1\}, \emptyset)$; the extra "0" in X_L didn't seem to make any difference.

A. If what I said is true, it will save us a lot of work, because each number (X_L, X_R) will behave in all $\leq$ relations exactly as if X_L were replaced by its largest element and X_R by its smallest. We won't have to consider any numbers in which X_L or X_R have two or more elements; ten of those twenty numbers in the list will be eliminated!

B. I'm not sure I follow you, but how on earth can we prove such a thing?

A. What we seem to need is something like this:

$$\text{if} \quad x \leq y \quad \text{and} \quad y \leq z, \quad \text{then} \quad x \leq z. \quad \text{(T1)}$$

I don't see that this follows immediately, although it is consistent with everything we know.

B. At any rate, it ought to be true, if Conway's numbers are to be at all decent. We could go ahead and assume it, but it would be neat to show once and for all that it was true, just by using Conway's rules.

A. Yes, and we'd be able to solve the Third Day puzzle without much more work. Let's see, how can it be proved . . . .

B. Blast these flies! Just when I'm trying to concentrate. Alice, can you—no, I guess I'll go for a little walk.

. . . .

Any progress?

A. No, I seem to be going in circles, and the $\not\geq$ versus $\leq$ is confusing. Everything is stated negatively and things get incredibly tangled up.

B. Maybe (T1) isn't true.

A. But it *has* to be true. Wait, that's it! We'll try to *disprove* it. And when we fail, the cause of our failure will be a proof!

B. Sounds good—it's always easier to prove something wrong than to prove it right.

A. Suppose we've got three numbers x, y, and z for which

$$x \leq y, \quad \text{and} \quad y \leq z, \quad \text{and} \quad x \not\leq z.$$

What does rule (2) tell us about "bad numbers" like this?

B. It says that

$$\begin{aligned} &X_L \not\geq y, \\ \text{and} \quad &x \not\geq Y_R, \\ \text{and} \quad &Y_L \not\geq z, \\ \text{and} \quad &y \not\geq Z_R, \end{aligned}$$

and then also $x \not\leq z$, which means what?

A. It means one of the two conditions fails. Either there is a number x_L in X_L for which $x_L \geq z$, or there is a number z_R in Z_R for which $x \geq z_R$. With all these facts about x, y, and z, we ought to be able to prove *something*.

B. Well, since x_L is in X_L, it can't be greater than or equal to y. Say it's less than y. But $y \leq z$, so x_L must be ... no, sorry, I can't use facts about numbers we haven't proved.

Going the other way, we know that $y \leq z$ and $z \leq x_L$ and $y \not\leq x_L$; so this gives us three more bad numbers, and we can get more facts again. But that looks hopelessly complicated.

A. Bill! You've got it.

B. Have I?

A. If (x, y, z) are three bad numbers, there are two possible cases.

Case 1, some $x_L \geq z$: Then (y, z, x_L) are three more bad numbers.

Case 2, some $z_R \leq x$: Then (z_R, x, y) are three more bad numbers.

B. But aren't you still going in circles? There's more and more bad numbers all over the place.

A. No, in each case the new bad numbers are *simpler* than the original ones; one of them was created earlier. We can't go on and on finding earlier and earlier sets of bad numbers, so there can't be any bad sets at all!

B. (brightening) Oho! What you're saying is this: Each number x was created on some day $d(x)$. If there are three bad numbers (x, y, z), for which the sum of their creation days is $d(x) + d(y) + d(z) = n$, then one of your two cases applies and gives three bad numbers whose day-sum is less than n. Those, in turn, will produce a set whose day-sum is still less and so on; but that's impossible since there are no three numbers whose day-sum is less than 3.

A. Right, the sum of the creation days is a nice way to express the proof. If there are no three bad numbers (x, y, z) whose day-sum is less than n, the two cases show that there are none whose day-sum equals n. I guess it's a proof by induction on the day-sum.

B. You and your fancy words. It's the *idea* that counts.

A. True; but we need a name for the idea, so we can apply it more easily next time.

B. Yes, I suppose there will be a next time . . .

Okay, I guess there's no reason for me to be uptight any more about the New Math jargon. You know it and I know it; we've just proved the *transitive law*.

A. (sigh) Not bad for two amateur mathematicians!

B. It was really your doing. I hereby proclaim that the transitive law (T1) shall be known henceforth as Alice's Theorem.

A. C'mon. I'm sure Conway discovered it long ago.

B. But does that make your efforts any less creative? I bet every great mathematician started by rediscovering a bunch of "well-known" results.

A. Gosh, let's not get carried away dreaming about greatness! Let's just have fun with this.

5

PROGRESS

B. I just thought of something. Could there possibly be two numbers that aren't related to each other at all? I mean

$$x \not\lessgtr y \quad \text{and} \quad y \not\lessgtr x,$$

like one of them is out of sight or in another dimension or something. It shouldn't happen, but how would we prove it?

A. I suppose we could try the same technique that worked before. If x and y are bad numbers in this sense, then either some $x_L \geq y$ or $x \geq$ some y_R.

B. Hmm. Suppose $y \leq x_L$. Then if $x_L \leq x$, we would have $y \leq x$ by our transitive law, and we have assumed that $y \nleq x$. So $x_L \nleq x$. In the other case, $y_R \leq x$, the same kind of figuring would show that $y \nleq y_R$.

A. Hey, that's very shrewd! All we have to do now to show that such a thing can't happen is prove something I've suspected for a long time. Every number x must lie between all the elements of its sets X_L and X_R. I mean,

$$X_L \leq x \qquad \text{and} \qquad x \leq X_R. \tag{T2}$$

B. That shouldn't be hard to prove. What does $x_L \nleq x$ say?

A. Either there is a number x_{LL} in X_{LL}, with $x_{LL} \geq x$, or else there is a number x_R in X_R with $x_L \geq x_R$. But the second case can't happen, by rule (1).

B. I *knew* we were going to use rule (1) sooner or later. But what can we do with x_{LL}? I don't like double subscripts.

A. Well, x_{LL} is an element of the left set of x_L. Since x_L was created earlier than x, we can at least assume that $x_{LL} \leq x_L$, by induction.

B. Lead on.

A. Let's see, $x_{LL} \leq x_L$ says that $x_{LLL} \nleq x_L$ and . . .

B. (interrupting) I don't want to look at this—your subscripts are getting worse.

A. You're a big help.

B. Look, I *am* helping, I'm telling you to keep away from those hairy subscripts!

A. But I . . . Okay, you're right, excuse me for going off on such a silly tangent. We have $x \leq x_{LL}$ and $x_{LL} \leq x_L$, so the transitive law tells us that $x \leq x_L$. This probably gets around the need for extra subscripts.

B. Aha, that does it. We can't have $x \leq x_L$, because that would mean $X_L \npreceq x_L$, which is impossible since x_L is one of the elements of X_L.

A. Good point, but how do you know that $x_L \leq x_L$.

B. What? You mean we've come this far and haven't even proved that a number is like itself? Incredible . . . there must be an easy proof.

A. Maybe you can see it, but I don't think it's obvious. At any rate, let's try to prove

$$x \leq x. \tag{T3}$$

This means that $X_L \npreceq x$ and $x \npreceq X_R$.

B. It's curiously like (T2). But uh-oh, here we are in the same spot again, trying to show that $x \leq x_L$ is impossible.

A. This time it's all right, Bill. Your argument shows that $x \leq x_L$ implies $x_L \npreceq x_L$, which is impossible by induction.

B. Beautiful! That means (T3) is true, so everything falls into place. We've got the "$X_L \leq x$" half of (T2) proved, and the other half must follow by the same argument, interchanging left and right everywhere.

A. And like we said before, (T2) is enough to prove that all numbers are related; in other words

$$\text{if} \quad x \nleq y \quad \text{then} \quad y \leq x. \tag{T4}$$

B. Right. Look, now we don't have to bother saying things so indirectly any more, since "$x \not\geq y$" is exactly the same as "x is less than y."

A. I see, it's the same as "x is less than or like y but not like y." We can now write

$$x < y$$

in place of $x \not\geq y$, and the original rules (1) and (2) look much nicer. I wonder why Conway didn't define things that way? Maybe it's because a third rule would be needed to define what "less than" means, and he probably wanted to keep down the number of rules.

B. I wonder if it's possible to have two different numbers which are like each other. I mean, can we have both $x \leq y$ and $x \geq y$ when X_L is not the same set as Y_L?

A. Sure, we saw something like that before lunch. Don't you remember, we found that $0 \leq y$ and $y \leq 0$ when $y = (\{-1\}, \emptyset)$. And I think $(\{0, 1\}, \emptyset)$ will turn out to be like $(\{1\}, \emptyset)$.

B. You're right. When $x \leq y$ and $x \geq y$, I guess x and y are effectively equal for all practical purposes, because the transitive law tells us that $x \leq z$ if and only if $y \leq z$. They're interchangeable.

A. Another thing, we've also got two more transitive laws, I mean

| if | $x < y$ | and | $y \leq z$ | then | $x < z$; | (T5) |
| if | $x \leq y$ | and | $y < z$ | then | $x < z$. | (T6) |

B. Very nice—in fact, these both follow immediately from (T1), if we consider "$x < y$" equivalent to "$x \not\geq y$". There's no need to use (T2), (T3), or (T4) in the proofs of (T5) and (T6).

32

A. You know, when you look over everything we've proved, it's really very pretty. It's amazing that so much flows out of Conway's two rules.

B. Alice, I'm seeing a new side of you today. You really put to rest the myth that women can't do mathematics.

A. Why, thank you, gallant knight!

B. I know it sounds crazy, but working on this creative stuff with you makes me feel like a stallion! You'd think so much brainwork would turn off any physical desires, but really—I haven't felt quite like this for a long time.

A. To tell the truth, neither have I.

B. Look at that sunset, just like in the poster we bought once. And look at that water.

A. (running) Let's go!

6 THE THIRD DAY

B. Boy, I never slept so well.

A. Me too. It's so great to wake up and be really awake, not just "coffee-awake."

B. Where were we yesterday, before we lost our heads and forgot all about mathematics?

A. (smiling) I think we had just proved that Conway's numbers behave like all little numbers should; they can be arranged in a line, from smallest to largest, with every number being greater than those to its left and less than all those on its right.

B. Did we really prove that?

A. Yes, anyway at least the unlike numbers keep in line, because of (T4). Every new number created must fall into place among the others.

B. Now it should be pretty easy for us to figure out what happened on the Third Day; those 20 × 20 calculations must be reduced 'way down. Our theorems (T2) and (T3) show that

$$⟨:\text{-}⟩ \,<\, \text{-} \,<\, ⟨\text{-}:\bullet⟩ \,<\, \bullet \,<\, ⟨\bullet:|⟩ \,<\, | \,<\, ⟨|:⟩$$

so seven of the numbers are placed already and it's just a matter of fitting the other ones in.

You know, now that it's getting easier, this is much more fun than a crossword puzzle.

A. We also know, for example, that

$$⟨\text{-}:|⟩$$

lies somewhere between ⬤ and |. Let's check it against the middle element 0.

B. Hmm, it's both $\leq$ and ≥ 0, so it must be like 0, according to rule (3). As I said yesterday, it's effectively equal to 0, so we might as well forget it. That's eight down and twelve to go.

A. Let's try to get rid of those ten cases where X_L or X_R have more than one element, like I tried to do yesterday morning. I had an idea during the night which might work. Suppose

$x = (X_L, X_R)$ is a number, and we take any other sets of numbers Y_L and Y_R, where

$$Y_L < x < Y_R.$$

Then I think it's true that x is like z, where

$$z = (Y_L \cup X_L, X_R \cup Y_R).$$

In other words, enlarging the sets X_L and X_R, by adding numbers on the appropriate sides, doesn't really change x.

B. Let's see, that sounds plausible. At any rate, z is a number, according to rule (1); it will be created sooner or later.

A. In order to show that $z \leq x$, we have to prove that

$$Y_L \cup X_L < x \qquad \text{and} \qquad z < X_R.$$

But that's easy, now, since we know that $Y_L < x$, $X_L < x$, and $z < X_R \cup Y_R$, by (T3).

B. And the same argument, interchanging left and right, shows that $x \leq z$. You're right, it's true:

$$\begin{aligned} &\text{if} \qquad Y_L < x < Y_R, \\ &\text{then} \qquad x \equiv (Y_L \cup X_L, X_R \cup Y_R). \end{aligned} \qquad \text{(T7)}$$

(I'm going to write "$x \equiv z$," meaning x is like z, I mean $x \leq z$ and $z \leq x$.)

A. That proves just what we want. For example,

$$\langle \text{-●:I} \rangle \equiv \langle \text{●:I} \rangle, \ \langle \text{:-●} \rangle \equiv \langle \text{:-} \rangle$$

and so on.

B. So we're left with only two cases: 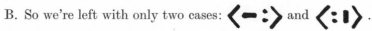 .

A. Actually, (T7) applies to both of these, too, with $x = 0$!

B. Cle-*ver*. So the Third Day is now completely analyzed; only those seven numbers we listed before are essentially different.

A. I wonder if the same thing won't work for the following days, too. Suppose the different numbers at the end of n days are

$$x_1 < x_2 < \cdots < x_m.$$

Then maybe the only new numbers created on the $(n + 1)$st day will be

$$(\emptyset, \{x_1\}), \quad (\{x_1\}, \{x_2\}), \quad \ldots, \quad (\{x_{m-1}\}, \{x_m\}), \quad (\{x_m\}, \emptyset).$$

B. Alice, you're wonderful! If we can prove this, it will solve infinitely many days in one swoop! You'll get ahead of the Creator himself.

A. But maybe we can't prove it.

B. Anyway let's try some special cases. Like, what if we had the number $(\{x_{i-1}\}, \{x_{i+1}\})$; it would have to be equal to one of the others.

A. Sure, it equals x_i, because of (T7). Look, each element of X_{iL} is $\leq x_{i-1}$, and each element of X_{iR} is $\geq x_{i+1}$. Therefore, by (T7), we have

$$x_i \equiv (\{x_{i-1}\} \cup X_{iL}, X_{iR} \cup \{x_{i+1}\}).$$

And again by (T7),

$$(\{x_{i-1}\}, \{x_{i+1}\}) \equiv (X_{iL} \cup \{x_{i-1}\}, \{x_{i+1}\} \cup X_{iR}).$$

By the transitive law, $x_i \equiv (\{x_{i-1}\}, \{x_{i+1}\})$.

B. (shaking his head) Incredible, Holmes!

A. Elementary, my dear Watson. One simply uses deduction.

B. Your subscripts aren't very nice, but I'll ignore it this time. What would you do with the number $(\{x_{i-1}\}, \{x_{j+1}\})$ if $i < j$?

A. (shrugging her shoulders) I was afraid you'd ask that. I don't know.

B. Your same argument would work beautifully if there was a number x where each element of X_L is $\leq x_{i-1}$ and each element of X_R is $\geq x_{j+1}$.

A. Yes, you're right, I hadn't noticed that. But all those elements $x_i, x_{i+1}, \ldots, x_j$ in between might interfere.

B. I suppose so ... No, I've got it! Let x be the one of $x_i, x_{i+1}, \ldots, x_j$ which was created *first*. Then X_L and X_R can't involve any of the others! So $(\{x_{i-1}\}, \{x_{j+1}\}) \equiv x$.

A. Allow me to give you a kiss for that.

. . . .

. . . .

B. (smiling) The problem isn't completely solved, yet; we have to consider numbers like $(\emptyset, \{x_{j+1}\})$ and $(\{x_{i-1}\}, \emptyset)$. But in the first case, we get the first-created number of $x_1, x_2, \ldots, x_j$, and in the second case it's the first-created number of $x_i, x_{i+1}, \ldots, x_m$.

A. What if the first-created number wasn't unique? I mean, what if more than one of the $x_i, \ldots, x_j$ were created on that earliest day?

B. Whoops ... No, it's okay, that can't happen, because the proof is still valid and it would show that the two numbers are both like each other, which is impossible.

A. Neato! You've solved the problem of all the days at once.

B. With your help. Let's see, on the fourth day there will be 8 new numbers, then on the fifth day there are 16 more, and so on.

A. Yes, after the nth day, exactly $2^n - 1$ different numbers will have been created.

B. You know, I don't think that guy Conway was so smart after all. I mean, he could have just given much simpler rules, with the same effect. There's no need to talk about sets of numbers, and all that nonsense; he simply would have to say that the new numbers are created between existing adjacent ones, or at the ends.

C. **Rubbish. Wait until you get to infinite sets.**

A. What was that? Did you hear something? It sounded like thunder.

B. I'm afraid we'll be getting into the monsoon season pretty soon.

7 DISCOVERY

A. Well, we've solved everything on that rock, but I can't help feeling there's still a lot missing.

B. What do you mean?

A. I mean, like we know what happened on the third day, four new numbers were created. But we don't know what Conway called them.

B. Well, one of the numbers was bigger than 1, so I suppose he called it "2." And another was between 0 and 1, so maybe he called it "$\frac{1}{2}$."

A. That's not really the point; what really bothers me is, why are they *numbers*? I mean, in order to be numbers you have to be added, subtracted, and that sort of thing.

B. (frowning) I see. You think Conway gave some more rules, in the broken-off part of the rock, which made the numbers numerical. All we have is a bunch of objects ordered neatly in a line, but we haven't got anything to do with them.

A. I don't think I'm clairvoyant enough to guess what he did—if he did do something.

B. That means we're stuck, unless we can find the missing part of that rock. And I don't remember where we found the first part.

A. Oh, I remember that, I was careful to note exactly where it was in case we ever wanted to go back.

B. What would I do without you? Come on, let's go!

A. Hey wait, don't you think we should have a little lunch first?

B. Right, I got so wrapped up in this I forgot all about food. Okay, let's grab a quick bite and then start digging.

. . . .

A. (digging) Oh, Bill, I'm afraid this isn't going to work. The dirt under the sand is so hard, we need special tools.

B. Yeah, just scraping away with this knife isn't getting us very far. Uh oh, here comes the rain, too. Should we dash back to camp?

A. Look, there's a cave over by that cliff. Let's wait out the storm in there. Hey, it's really pouring!

. . . .

B. Sure is dark in here. Ouch! I stubbed my toe on something. Of all the . . .

A. Bill! You've found it! You stubbed your toe on the other part of the Conway Stone!

B. (wincing) Migosh, it look's like you're right. Talk about fate! But my toe isn't as pleased about it as the rest of me is.

A. Can you read it, Bill? Is it really the piece we want, or is it something else entirely?

B. It's too dark in here to see much. Help me drag it out in the rain, the water will wash the dust off and . . .

Yup, I can make out the words "Conway" and "number," so it *must* be what we're looking for.

A. Oh, good, we'll have plenty to work on. We're saved!

B. The info we need is here all right. But I'm going back in the cave, it can't keep raining this hard for very long.

A. (following) Right, we're getting drenched.

. . . .

B. I wonder why this mathematics is so exciting now, when it was so dull in school. Do you remember old Professor Landau's lectures? I used to really hate that class: Theorem, proof, lemma, remark, theorem, proof, what a total drag.

A. Yes, I remember having a tough time staying awake. But look—wouldn't *our* beautiful discoveries be just about the same?

B. True. I've got this mad urge to get up before a class and present our results: Theorem, proof, lemma, remark. I'd make it so slick, nobody would be able to guess how we did it, and everyone would be *so* impressed.

A. Or bored.

B. Yes, there's that. I guess the excitement and the beauty comes in the discovery, not the hearing.

A. But it *is* beautiful. And I enjoyed hearing your discoveries almost as much as making my own. So what's the real difference?

B. I guess you're right, at that. I was able to really appreciate what *you* did, because I had already been struggling with the same problem myself.

A. It was dull before, because we weren't involved at all; we were just being told to absorb what somebody else did, and for all we knew there was nothing special about it.

B. From now on whenever I read a math book, I'm going to try to figure out by myself how everything was done, before looking at the solution. Even if I don't figure it out, I think I'll be able to see the beauty of a proof then.

A. And I think we should also try to guess what theorems are coming up; or at least, to figure out how and why anybody would try to prove such theorems in the first place. We should imagine ourselves in the discoverer's place. The creative part is really more interesting than the deductive part. Instead of concentrating just on finding good answers to questions, it's more important to learn how to find good questions!

B. You've got something there. I wish our teachers would give us problems like, "Find something interesting about x," instead of "Prove x."

A. Exactly. But teachers are so conservative, they'd be afraid of scaring off the "grind" type of students who obediently and mechanically do all the homework. Besides, they wouldn't like the extra work of grading the answers to nondirected questions.

The traditional way is to put off all creative aspects until the last part of graduate school. For seventeen or more years, a student is taught examsmanship, then suddenly after passing enough exams in graduate school he's told to do something original.

B. Right. I doubt if many of the really original students have stuck around that long.

A. Oh, I don't know, maybe they're original enough to find a way to enjoy the system. Like putting themselves into the subject, as we were saying. That would make the traditional college courses tolerable, maybe even fun.

B. You always were an optimist. I'm afraid you're painting too rosy a picture. But look, the rain has stopped, let's lug this rock back to camp and see what it says.

ADDITION

A. The two pieces fit pretty well, it looks like we've got almost the whole message. What does it say?

B. This part is a little harder to figure out, there are some obscure words, but I think it goes like this:

> ... day. And Conway said, "Let the numbers be added to each other in this wise: The left set of the sum of two numbers shall be the sums of all left parts of each number

with the other; and in like manner the right set shall be from the right parts, each according to his kind." Conway proved that every number plus zero is unchanged, and he saw that addition was good. And the evening and the morning were the third day.

And Conway said, "Let the negative of a number have as its sets the negatives of the number's opposite sets; and let subtraction be addition of the negative." And it was so. Conway proved that subtraction was the inverse of addition, and this was very good. And the evening and the morning were the fourth day.

And Conway said to the numbers, "Be fruitful and multiply. Let part of one number be multiplied by another and added to the product of the first number by part of the other, and let the product of the parts be subtracted. This shall be done in all possible ways, yielding a number in the left set of the product when the parts are of the same kind, but in the right set when they are of opposite kinds." Conway proved that every number times one is unchanged. And the evening and the morning were the fifth day.

And behold! When the numbers had been created for infinitely many days, the universe itself appeared. And the evening and the morning were $\aleph$ day.

And Conway looked over all the rules he had made for numbers, and saw that they were very, very good. And he commanded them to be for signs, and series, and quotients, and roots.

Then there sprang up an infinite number less than infinity. And infinities of days brought forth multiple orders of infinities.

That's the whole bit.

A. What a weird ending. And what do you mean "aleph day"?

B. Well, aleph is a Hebrew letter and it's just standing there by itself, look: ℵ. It seems to mean infinity. Let's face it, it's heavy stuff and it's not going to be easy to figure out what this means.

A. Can you write it all down while I fix supper? It's too much for me to keep in my head, and I can't read it.

B. Okay, that'll help me get it clearer in my own mind too.

. . . .

A. It's curious that the four numbers created on the third day aren't mentioned. I still wonder what Conway called them.

B. Maybe if we try the rules for addition and subtraction we could figure out what the numbers are.

A. Yeah, *if* we can figure out those rules for addition and subtraction. Let's see if we can put the addition rule into symbolic form, in order to see what it means . . . I suppose "its own kind" must signify that left goes with left, and right with right. What do you think of this:

$$x + y = ((X_L + y) \cup (Y_L + x), (Y_R + x) \cup (X_R + y)). \tag{3}$$

B. Looks horrible. What does *your* rule mean?

A. To get the left set of $x + y$, you take all numbers of the form $x_L + y$, where x_L is in X_L, and also all numbers $y_L + x$ where y_L is in Y_L. The right set is from the right parts, "in like manner."

B. I see, a "left part" of x is an element of X_L. Your symbolic definition certainly seems consistent with the prose one.

A. And it makes sense too, because each $x_L + y$ and $x + y_L$ ought to be less than $x + y$.

B. Okay, I'm willing to try it and see how it works. I see you've called it rule (3).

A. Now after the third day, we know that there are seven numbers, which we might call 0, 1, -1, a, b, c, and d.

B. No, I have an idea that we can use left-right symmetry and call them

$$-a < -1 < -b < 0 < b < 1 < a,$$

where

$$-a = \langle : - \rangle \qquad \langle \mathbf{I} : \rangle = a$$

$$-1 = \blacksquare = \langle : \bullet \rangle \qquad \langle \bullet : \rangle = \mathbf{I} = 1$$

$$-b = \langle - : \bullet \rangle \qquad \langle \bullet : \mathbf{I} \rangle = b$$

$$0 = \langle : \rangle = \bullet$$

A. Brilliant! You must be right, because Conway's next rule is

$$-x = (-X_R, -X_L), \tag{4}$$

B. So it is! Okay—now we can start adding these numbers. Like, what's $1 + 1$, according to rule (3)?

A. You work on that, and I'll work on $1 + a$.

B. Okay, I get $(\{0 + 1, 0 + 1\}, \emptyset)$. And $0 + 1$ is $(\{0 + 0\}, \emptyset)$, $0 + 0$ is $(\emptyset, \emptyset) = 0$. Everything fits together, making $1 + 1 = (\{1\}, \emptyset) = a$. Just as we thought, a must be 2!

A. Congratulations on coming up with the world's longest proof that $1 + 1$ is 2.

B. Have you ever seen a shorter proof?

A. Not really. Look, your calculations help me too. I get $1 + 2 = (\{2\}, \emptyset)$, a number that isn't created until the fourth day.

B. I suggest we call it "3."

A. Bravo. So rule (3) is working; let's check if b is $\frac{1}{2}$ by calculating $b + b \ldots$

B. Hmm, that's odd, it comes out to $(\{b\}, \{b + 1\})$, which hasn't been created yet.

A. And $b + 1$ is $(\{b, 1\}, \{2\})$, which is like $(\{1\}, \{2\})$, which is created on the fourth day. So $b + b$ appears on the *fifth* day.

B. Don't tell me $b + b$ is going to be equal to *another* number we don't know the name of.

A. Are we stuck?

B. We worked out a theory that tells us how to calculate all numbers that are created, so we *should* be able to do this. Let's make a table for the first four days.

A. Oh, Bill, that's too much work.

B. No, it's a simple pattern really. Look:

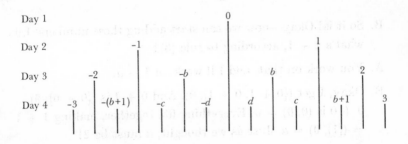

A. Oh I see, so $b + b$ is $(b, b + 1)$, which is formed from *non-adjacent* numbers ... And our theory says it is the *earliest-created* number between them.

B. (beaming) And that's 1, because 1 makes the scene before c.

A. So b is $\frac{1}{2}$ after all, although its numerical value wasn't established until two days later. It's amazing what can be proved from those few rules—they all hang together so tightly, it boggles the mind.

B. I'll bet d is $\frac{1}{3}$ and c is $\frac{2}{3}$.

A. But the sun is going down. Let's sleep on it, Bill; we've got lots of time and I'm really drained.

B. (muttering) $d + c = $ ... Oh, all right. G'night.

A. Are you awake already?

B. What a miserable night! I kept tossing and turning, and my mind was racing in circles. I dreamed I was proving things and making logical deductions, but when I woke up they were all foolishness.

A. Maybe this mathematics isn't good for us after all. We were so happy yesterday, but—

B. (interrupting) Yeah, yesterday we were high on math, but today it's turning sour. I can't get it out of my system, we've *got* to get more results before I can rest. Where's that pencil?

A. Bill, you need some breakfast. There are some apricots and figs over there.

B. Okay, but I've gotta get right to work.

A. Actually I'm curious to see what happens too, but promise me one thing.

B. What?

A. We'll only work on addition and subtraction today; *not* multiplication. We won't even *look* at that other part of the tablet until later.

B. Agreed. I'm almost willing to postpone the multiplication indefinitely, since it looks awfully complicated.

A. (kissing him) Okay, now relax.

B. (stretching) You're so good to me, Alice.

A. That's better. Now I was thinking last night about how you solved the problem about all the numbers yesterday morning. I think it's an important principle that we ought to write down as a theorem. I mean:

> Given any number y, if x is the first number created with the property that $Y_L < x$ and $x < Y_R$, then $x \equiv y$. (T8)

B. Hmm, I guess that *is* what we proved. Let's see if we can reconstruct the proof, in this new symbolism. As I recall we constructed the number $z = (Y_L \cup X_L, X_R \cup Y_R)$, and then we had $x \equiv z$ by (T7). On the other hand, no element x_L of X_L satisfies $Y_L < x_L$, since x_L was created before x; therefore each x_L is $\leq$ some y_L, by (T4). Thus $X_L < y$, and similarly, $y < X_R$. So $y \equiv z$ by (T7).

It's pretty easy to work out the proof now that we have all this ammunition to work with.

A. The nice thing about (T8) is that it makes the calculation we did last night much easier. Like when we were calculating $b + b = (\{b\}, \{b + 1\})$, we could have seen immediately that 1 is the first number created between $\{b\}$ and $\{b + 1\}$.

B. Hey, let me try that on $c + c$: It's the first number created between $b + c$ and $1 + c$. Well, it must be $b + 1$, I mean $1\frac{1}{2}$, so c is $\frac{3}{4}$.

That's a surprise, I thought it would be $\frac{2}{3}$.

A. And d is $\frac{1}{4}$.

B. Right.

A. I think the general pattern is becoming clear now: After four days the numbers ≥ 0 are

$$0, \tfrac{1}{4}, \tfrac{1}{2}, \tfrac{3}{4}, 1, \tfrac{3}{2}, 2, 3$$

and after five days they will probably be—

B. (interrupting)

$$0, \tfrac{1}{8}, \tfrac{1}{4}, \tfrac{3}{8}, \tfrac{1}{2}, \tfrac{5}{8}, \tfrac{3}{4}, \tfrac{7}{8}, 1, \tfrac{5}{4}, \tfrac{3}{2}, \tfrac{7}{4}, 2, \tfrac{5}{2}, 3, 4.$$

A. Exactly. Can you prove it?

B. . . .

Yes, but not so easily as I thought. For example, to figure out the value of $f = (\{\tfrac{3}{2}\}, \{2\})$, which turned out to be $\frac{7}{4}$, I calculated $f + f$. This is the first number created between 3 and 4, and I had to "look ahead" to see that it was $\frac{7}{2}$. I'm convinced we have the right general pattern, but it would be nice to have a proof.

A. On the fourth day we calculated $\frac{3}{2}$ by knowing that it was $1 + \frac{1}{2}$, *not* by trying $\frac{3}{2} + \frac{3}{2}$. Maybe adding 1 will do the trick.

B. Let's see . . . According to the definition, rule (3),

$$1 + x = ((1 + X_L) \cup \{x\}, 1 + X_R),$$

assuming that $0 + x = x$. In fact, isn't it true that . . . sure, for positive numbers we can always choose X_L so that $1 + X_L$ has an element $\geq x$, so it simplifies to

$$1 + x = (1 + X_L, 1 + X_R)$$

in this case.

A. That's it, Bill! Look at the last eight numbers on the fifth day, they are just one greater than the eight numbers on the fourth day.

B. A perfect fit. Now all we have to do is prove the pattern for the numbers x between 0 and 1 . . . but that can always be done by looking at $x + x$, which will be less than 2!

A. Yes, now I'm sure we've got the right pattern.

B. What a load off my mind. I don't even feel the need to formalize the proof now; I *know* it's right.

A. I wonder if our rule for $1 + x$ isn't a special case of a more general rule. Like, isn't

$$y + x = (y + X_L, y + X_R)?$$

That would be much simpler than Conway's complicated rule.

B. Sounds logical, since adding y should "shift" things over by y units. Whoops, no, take $x = 1$; that would say $y + 1$ is $(\{y\}, \emptyset)$, which fails when y is $\frac{1}{2}$.

A. Sorry. In fact, your rule for $1 + x$ doesn't work when $x = 0$ either.

B. Right, I proved it only when x is positive.

A. I think we ought to look at rule (3), the addition rule, more closely and see what can be proved in general from it. All we've got are *names* for the numbers. These names must be correct if Conway's numbers behave like actual numbers, but we don't know that Conway's rules are really the same. Besides, I think it's fun to derive a whole bunch of things from just a few basic rules.

B. Let's see. In the first place, addition is obviously what we might call commutative, I mean

$$x + y = y + x. \tag{T9}$$

A. True. Now let's prove what Conway claimed, that

$$x + 0 = x. \tag{T10}$$

B. The rule says that

$$x + 0 = (X_L + 0, X_R + 0).$$

So all we do is a "day of creation" induction argument, again; we can assume that $X_L + 0$ is the same as X_L, and $X_R + 0$ is X_R, since all those numbers were created before x. Q.E.D.

A. Haven't we proved that $x + 0 \equiv x$, not $= x$?

B. You're a nit-picker, you are. I'll change (T10) if you want me to, since it really won't make any difference. But actually doesn't the proof actually show that $x + 0$ is identically the same pair of sets as x?

A. Excuse me again. You're right.

B. That's ten theorems. Shall we try for more while we're hot?

10

A. How about the associative law,

$$(x + y) + z = x + (y + z). \tag{T11}$$

B. Oh, I doubt if we'll need that; it didn't come up in the calculations. But I suppose it won't hurt to try it, since my math teachers always used to think it was such a great thing.

One associative law, coming right up. Can you work out the definition?

A.
$$(x + y) + z = (((X_L + y) + z) \cup ((Y_L + x) + z)$$
$$\cup (Z_L + (x + y)), ((X_R + y) + z)$$
$$\cup ((Y_R + x) + z) \cup (Z_R + (x + y)))$$

$$x + (y + z) = ((X_L + (y + z)) \cup ((Y_L + z) + x)$$
$$\cup ((Z_L + y) + x), (X_R + (y + z))$$
$$\cup ((Y_R + z) + x) \cup ((Z_R + y) + x)).$$

B. You're really good at these hairy formulas. But how can such monstrous things be proved equal?

A. It's not hard, just using a day-sum argument on (x, y, z) as we did before. See, $(X_L + y) + z = X_L + (y + z)$ because (x_L, y, z) has a smaller day-sum than (x, y, z), and we can induct on that. The same for the other five sets, using the commutative law in some cases.

B. Congratulations! Another Q.E.D., and another proof of $=$ instead of $\equiv$.

A. That $\equiv$ worries me a little, Bill. We showed that we could substitute like elements for like elements, with respect to $<$ and $\leq$, but don't we have to verify this also for addition? I mean,

$$\text{if} \quad x \equiv y, \quad \text{then} \quad x + z \equiv y + z. \tag{T12}$$

B. I suppose so, otherwise we wouldn't strictly be allowed to make the simplifications we've been making in our names for the numbers. As long as we're proving things, we might as well do it right.

A. In fact, we might as well prove a stronger statement,

$$\text{if} \quad x \leq y, \quad \text{then} \quad x + z \leq y + z, \quad \text{(T13)}$$

because this will immediately prove (T12).

B. I see, because $x \equiv y$ if and only if $x \leq y$ and $y \leq x$. Also (T13) looks like it will be useful. Shouldn't we also prove more, I mean

$$\text{if} \quad x \leq y \quad \text{and} \quad w \leq z,$$
$$\text{then} \quad x + w \leq y + z?$$

A. Oh, that follows from (T13), since $x + w \leq y + w = w + y \leq z + y = y + z$.

B. Okay, that's good, because (T13) is simpler. Well, you're the expert on formulas, what is (T13) equivalent to?

A. Given that $X_L < y$ and $x < Y_R$, we must prove that $X_L + z < y + z$, $Z_L + x < y + z$, $x + z < Y_R + z$, and $x + z < Z_R + y$.

B. Another day-sum induction, eh? Really, these are getting too easy.

A. Not quite so easy, this time. I'm afraid the induction will only give us $X_L + z \leq y + z$, and so on; it's conceivable that $x_L < y$ but $x_L + z \equiv y + z$.

B. Oh yeah. That's interesting. What we need is the converse,

$$\text{if} \quad x + z \leq y + z \quad \text{then} \quad x \leq y. \quad \text{(T14)}$$

A. Brilliant! The converse is equivalent to this: Given that $X_L + z < y + z$, $Z_L + x < y + z$, $x + z < Y_R + z$, and $x + z < Z_R + y$, prove that $X_L < y$ and $x < Y_R$.

B. Hmm. The converse would go through by induction—except that we might have a case with, say, $x_L + z < y + z$ but $x_L \equiv y$. Such cases would be ruled out by (T13), but . . .

A. But we need (T13) to prove (T14), and (T14) to prove (T13). And (T13) to prove (T12).

B. We're going around in circles again.

A. Ah, but there's a way out, we'll prove them *both* together! We can prove the combined statement "(T13) and (T14)" by induction on the day-sum of (x, y, z)!

B. (glowing) Alice, you're a genius! An absolutely gorgeous, tantalizing genius!

A. Not so fast, we've still got work to do. We had better show that

$$x - x \equiv 0. \tag{T15}$$

B. What's that minus sign? We never wrote down Conway's rule for subtraction.

A. $\qquad x - y = x + (-y). \tag{5}$

B. I notice you put the $\equiv$ in (T15); okay, it's clear that $x + (-x)$ won't be identically equal to 0, I mean with empty left and right sets, unless x is 0.

A. Rules (3), (4), and (5) say that (T15) is equivalent to this:

$$((X_L + (-x)) \cup ((-X_R) + x),$$
$$(X_R + (-x)) \cup ((-X_L) + x)) \equiv 0.$$

B. Uh oh, it looks hard. How do we show something $\equiv 0$ anyway? . . . By (T8), $y \equiv 0$ if and only if $Y_L < 0$ and $0 < Y_R$, since 0 was the first created number of all.

A. The same statement also follows immediately from rule (2); I mean, $y \leq 0$ if and only if $Y_L < 0$ and $0 \leq y$ if and only if $0 < Y_R$. So now what we have to prove is

$$x_L + (-x) < 0, \quad \text{and} \quad (-x_R) + x < 0,$$
$$\text{and } x_R + (-x) > 0, \quad \text{and} \quad (-x_L) + x > 0,$$

for all x_L in X_L and all x_R in X_R.

B. Hmm. Aren't we allowed to assume that $x_L + (-x_L) \equiv 0$ and $x_R + (-x_R) \equiv 0$?

A. Yes, since we can be proving (T15) by induction.

B. Then I've got it! If $x_L + (-x)$ were ≥ 0, then $(-X)_R + x_L$ would be > 0, by definition. But $(-X)_R$ is $-(X_L)$, which contains $-x_L$, and $(-x_L) + x_L$ is not > 0. Therefore $x_L + (-x)$ must be < 0, and the same technique works for the other cases too.

A. Bravo! That settles (T15).

B. What next?

A. How about this?

$$-(-x) = x. \tag{T16}$$

B. Sssss. That's trivial. Next?

A. All I can think of is Conway's theorem,

$$(x + y) - y \equiv x. \tag{T17}$$

B. What's that equivalent to?

A. It's a real mess . . . Can't we prove things without going back to the definitions each time?

B. Aha! Yes, it almost falls out by itself:

$$
\begin{aligned}
(x + y) - y &= (x + y) + (-y) && \text{by (5)} \\
&= x + (y + (-y)) && \text{by (T11)} \\
&= x + (y - y) && \text{by (5)} \\
&\equiv x + 0 && \text{by (T12) and (T15)} \\
&= x. && \text{by (T10)}
\end{aligned}
$$

We've built up quite a pile of useful results—even the associative law has come in handy. Thanks for suggesting it against my better judgment.

A. Okay, we've probably exhausted the possibilities of addition, negation, and subtraction. There are some more things we could probably prove, like

$$-(x + y) = (-x) + (-y), \tag{T18}$$

$$\text{if} \quad x \leq y, \quad \text{then} \quad -y \leq -x, \tag{T19}$$

but I don't think they involve any new ideas; so there's little point in proving them unless we need 'em.

B. Nineteen theorems, from just a few primitive rules.

A. Now you must remember your promise: This afternoon we take a vacation from mathematics, without looking at the rest of the stone again. I don't want that horrible multiplication jazz to rob you of any more sleep.

B. We've done a good day's work, anyhow—all the problems are resolved. Look, the tide's just right again. Okay—the last one into the water has to cook supper!

11 THE PROPOSAL

A. That sure was a good supper you cooked.

B. (lying down beside her) Mostly because of the fresh fish you caught.

What are you thinking about now?

A. (blushing) Well, actually I was wondering what would happen if I got pregnant.

B. You mean, here we are, near the Fertile Crescent, and . . . ?

A. Very funny. And after all our work to prove that $1 + 1 = 2$ we'll discover that $1 + 1 = 3$.

B. Okay, you win, no more jokes. But come to think of it, Conway's rules for numbers are like copulation, I mean the left set meeting the right set, . . .

A. You've got just one thing—no, two things—on your mind. But seriously, what would we do if I really were pregnant?

B. Well, I've been thinking we'd better go back home pretty soon anyway; our money's running out, and the weather is going to get bad.

Actually, I really want to marry you in any case, whether you're pregnant or not. If you'll have me, of course.

A. That's just what I feel too. This trip has proved that we're ready for a permanent relationship.

I wonder . . . When our children grow up, will we teach them our theory of numbers?

B. No, it would be more fun for them to discover it for themselves.

A. But people can't discover *everything* for themselves, there has to be some balance.

B. Well, isn't all learning really a process of self-discovery? Don't the best teachers help their students to think on their own?

A. In a way, yes. Whew, we're getting philosophical.

B. I still can't get over how great I feel when I'm doing this crazy mathematics; it really turns me on right now, but I used to hate it.

A. Yes, I've been high on it, too. I think it's a lot better than drugs; I mean, the brain can stimulate itself naturally.

B. And it was kind of an aphrodisiac, besides.

A. (gazing at the stars) One nice thing about pure mathematics—the things we proved today will never be good for anything, so nobody will be able to use them to make bombs or stuff like that.

B. Right. But we can't be in an ivory tower all the time, either. There are lots of problems in the world, and the right kind of math might help to solve them. You know, we've been away from newspapers for so long, we've forgotten all the problems.

A. Yeah, sometimes I feel guilty about that . . . .

Maybe the right kind of mathematics would help solve some of these problems, but I'm worried that it could also be misused.

B. That's the paradox, and the dilemma. Nothing can be done without tools, but tools can be used for bad things as well as good. If we stop creating things, because they might be harmful in the wrong hands, then we also stop doing useful things.

A. Okay, I grant you that pure mathematics isn't the answer to everything. But are you going to abolish it entirely just because it doesn't solve the world's problems?

B. Oh no, don't misunderstand me. These past few days have shown me that pure mathematics is beautiful—it's an art form like poetry or painting or music, and it turns us on. Our natural curiosity has to be satisfied. It would destroy us if we couldn't have some fun, even in the midst of adversity.

A. Bill, it's good to talk with you like this.

B. I'm enjoying it too. It makes me feel closer to you, and sort of peaceful.

12   DISASTER

B. Are you awake already?

A. About an hour ago I woke up and realized that there's a big, gaping hole in what we thought we proved yesterday.

B. No!

A. Yes, I'm afraid so. We forgot to prove that $x + y$ is a *number*.

B. You're kidding. Of course it's a number, it's the sum of two numbers! Oh, wait, I see . . . we have to check that rule (1) is satisfied.

A. Yes, the definition of addition isn't legitimate unless we can prove that $X_L + y < X_R + y$, and $X_L + y < Y_R + x$, and $Y_L + x < X_R + y$, and $Y_L + x < Y_R + x$.

B. These would follow from (T13) and (T14), but . . . I see your point, we proved (T13) and (T14) assuming that the sum of two numbers is a number. How did you ever think of this problem?

A. Well, that's kind of interesting. I was wondering what would happen if we defined addition like this:

$$x \oplus y = (X_L \oplus Y_L, X_R \oplus Y_R).$$

I called this $\oplus$ because it wasn't obviously going to come out the same as $+$. But it was pretty easy to see that $\oplus$ was a commutative and associative operation, so I wanted to see what it turned out to be.

B. I see; the sum of x and y lies between $X_L + Y_L$ and $X_R + Y_R$, so this definition might turn out to be simpler than Conway's.

A. But my hopes were soon dashed, when I discovered that

$$0 \oplus x = 0$$

for all x.

B. Ouch! Maybe $\oplus$ means multiplication?

A. Then I proved that $1 \oplus x = 1$ for all $x > 0$, and $2 \oplus x = 2$ for all $x > 1$, and $3 \oplus x = 3$ for all $x > 2$, and . . . .

B. I see. For all positive integers m and n, $m \oplus n$ is the *minimum* of m and n. That's commutative and associative, all right. So your $\oplus$ operation *did* turn out to be interesting.

A. Yes, and $\frac{1}{2} \oplus \frac{1}{2} = \frac{1}{2}$. But then I tried $(-\frac{1}{2}) \oplus \frac{1}{2}$, and I was stopped cold.

B. You mean . . . ? I see, $(-\frac{1}{2}) \oplus \frac{1}{2} = (\{(-1) \oplus 0\}, \{0 \oplus 1\})$, which is $(\{0\}, \{0\})$.

A. And that's *not* a number. It breaks rule (1).

B. So your definition of $\oplus$ wasn't legit.

A. And I realized that you can't just go making arbitrary definitions; they have to be proved consistent with the other rules too. Another problem with $\oplus$ was, for example, that $(\{-1\}, \emptyset) \equiv 0$ but $(\{-1\}, \emptyset) \oplus 1 \not\equiv 0 \oplus 1$.

B. Okay, $\oplus$ is out, but I suppose we can fix up the *real* definition of $+$.

A. I don't know; what I've just told you is as far as I got. Except that I thought about *pseudo-numbers*.

B. Pseudo-numbers?

A. Suppose we form (X_L, X_R) when X_L is not necessarily $< X_R$. Then rule (2) can still be used to define the $\leq$ relation between such pseudo-numbers.

B. I see . . . like $(\{1\}, \{0\})$ turns out to be less than 2.

A. Right. And I just noticed that our proof of the *transitive* law (T1) didn't use the $\not\geq$ part of rule (1), so that law holds for pseudo-numbers too.

B. Yes, I remember saying that the full rule (1) wasn't used until (T2). That seems like a long time ago.

A. Now get ready for a shock. The pseudo-number $(\{1\}, \{0\})$ is neither ≤ 0 nor ≥ 0!

B. Far out!

A. Yes, I think I can prove that $(\{1\}, \{0\})$ is $\leq$ a number y if and only if $y > 1$, and it is $\geq$ a number x if and only if $x < 0$. It's not related at all to any numbers between 0 and 1.

B. Where's the pencil? I want to check that out . . . I think you're right. This is fun, we're proving things about quantities that don't even exist.

A. Well, do pseudo-numbers exist any less than Conway's numbers? What you mean is, we're proving things about quantities that are purely conceptual, without real-world counterparts as aids to understanding . . . Remember that $\sqrt{-1}$ was once thought of as an imaginary number, and $\sqrt{2}$ wasn't even thought to be "rational."

B. Conway's rule for adding normal numbers also gives us a way to add pseudo-numbers. I wonder what this leads to? If $x = (\{1\}, \{0\})$, then $1 + x$ is . . . $(\{2\}, \{1\})$.

A. And $x + x$ is $(\{1 + x\}, \{x\})$, a second-order pseudo-number.

B. Pure mathematics is a real mind-expander.

   But did you notice that $(\{1\}, \{0\})$ isn't even $\leq$ itself?

A. Let's see, $x \leq x$ means that $X_L < x < X_R$, so this could only be true if $X_L < X_R$.

   No, wait, we aren't allowed to use "$<$" in place of "$\not\geq$" for pseudo-numbers, since (T4) isn't true in general. We have to go back to the original rule (2), which says that $x \leq x$ if and only if $X_L \not\geq x$ and $x \not\geq X_R$. So $(\{1\}, \{0\})$ *is* $\leq$ itself after all.

B. *Touché!* I'm glad I was wrong, since every x ought to be like itself, even when it's a pseudo-number.

A. Maybe there is some more complicated pseudo-number that isn't $\leq$ itself. It's hard to visualize, because the sets X_L and X_R might include pseudo-numbers too.

B. Let's look back at our proof of (T3) and see if it breaks down.

A. Good idea ... Hey, the same proof goes through for all pseudo-numbers: x is *always* like x.

B. This is great but I'm afraid it's taking us away from the main problem, whether or not $+$ is well-defined.

A. Well, our proofs that $x + y = y + x$, $x + 0 = x$, and even the associative law, work for pseudo-numbers as well as numbers. If the inequality theorems (T13) and (T14) also go through for pseudo-numbers, then $+$ will be well-defined.

B. I see, that's beautiful! So far we've established (T1), (T3), (T5), (T6), (T9), (T10), (T11) for all pseudo-numbers. Let's look at (T13) again.

A. But I'm afraid ... uh, oh. Bill! We were too gullible yesterday in our acceptance of that day-sum proof for (T13) and (T14); it was too good to be true.

B. What do you mean?

A. We were proving that $Z_L + x < y + z$ by induction, right? Well, to get this it takes two steps, first $Z_L + x \leq Z_L + y$ and then $Z_L + y < z + y$. Induction gives us the first part all right, but the second part involves (z_L, z, y), which might have a *larger* day-sum than (x, y, z).

B. So we really blew it. Conway would be ashamed of us.

A. Good thing we didn't see this yesterday, or it would have spoiled our day.

B. I guess it's back to the drawing boards ... but hey, we've *gotta* eat some breakfast.

13 RECOVERY

A. We've missed lunch, Bill.

B. (pacing the ground) Have we? This stupid problem is driving me up the wall.

A. Just staring at this paper isn't helping us any, either. We need a break; maybe if we ate something—

B. What we really need is a new idea. Gimme an idea, Alice.

A. (beginning to eat) Well, when we were going around in circles like this before, how did we break out? The main thing was to use induction, I mean to show that the proof in one case depended on the truth in a *previous* case, which depended on a still previous case, and so on, where the chain must eventually terminate.

B. Like our day-sum argument.

A. Right. The other way we broke the circle was by proving *more* than we first thought we needed. I mean, in order to keep the induction going, we had to keep proving several things simultaneously.

B. Like when you combined (T13) and (T14). Okay, Alice, right after lunch I'm going to sit down and write out the total picture, everything we need to prove, and perhaps even more. And I'm going to try and prove everything simultaneously by induction. The old battering-ram approach. If that doesn't work, nothing will.

A. That sounds hard but it's probably the best way. Here, have some oat cakes.

. . . .

B. Okay, here we go. We want to prove three things about numbers, and they all seem to depend on one another.

> I. $x + y$ is a number.
> II. if $x \leq y$, then $x + z \leq y + z$.
> III. if $x + z \leq y + z$, then $x \leq y$.

Now if I'm not mistaken, the proof of I(x, y) will follow if we have previously proved

84

$$I(X_L, y), \qquad I(x, Y_L), \qquad I(X_R, y), \qquad I(x, Y_R),$$
$$III(X_R, X_L, y),$$
$$III(x, X_L, y), \qquad II(y, Y_R, x),$$
$$III(y, Y_L, x), \qquad II(x, X_R, y),$$
$$III(Y_R, Y_L, x).$$

For example, we have to prove among other things that $X_L + y < Y_R + x$. In other words, for all x_L in X_L and y_R in Y_R we should have previously established that $x_L + y < y_R + x$. Now $III(x, x_L, y)$ and (T3) show that $x_L + y < x + y$, and $II(y, y_R, x)$ shows that $y + x \le y_R + x$. Right?

A. It looks good; except I don't see why you included those first four, $I(X_L, y)$ through $I(x, Y_R)$. I mean, even if $x_L + y$ wasn't a number, that wouldn't matter; all we really need to know is that x_L and y themselves are numbers. After all, $<$ and $\le$ are defined for pseudo-numbers, and the transitive laws work too.

B. No, rule (1) says that elements of the left part like $x_L + y$ have to be numbers. Anyway it doesn't really matter, because if we're proving $I(x, y)$ we can assume $I(x_L, y)$ and so on for free; induction takes care of them.

A. It's complicated, but keep going, this looks promising.

B. This approach *has* to work or we're sunk. Okay, the proof of $II(x, y, z)$, namely (T13), will follow if we have previously proved

$$III(y, X_L, z),$$
$$II(x, y, Z_L), \qquad III(z, Z_L, y),$$
$$III(Y_R, x, z),$$
$$II(x, y, Z_R), \qquad III(Z_R, z, x).$$

That's curious—this one really *doesn't* require $I(x, y)$. How come we thought we'd have to prove that the sum of two numbers is a number, before proving (T13)?

A. That was before we knew much about pseudo-numbers. It's strange how a fixed idea will remain as a mental block!

Remember? This was the first reason we said it was going to be hard to prove $x + y$ is a number, because we thought (T13) depended on this. After learning that pseudo-numbers satisfy the transitive laws, we forgot to reconsider the original source of trouble.

B. So at least this big picture method is getting us somewhere, if only because it helps organize our thoughts.

Now it's two down and one to go. The proof of $III(x, y, z)$ depends on knowing

$$II(X_L, y, z),$$
$$II(x, Y_R, z).$$

A. Again, $I(x, y)$ wasn't required. So we can simply prove (T13) and (T14) without worrying whether or not $x + y$ is a number.

B. I see—then later, $x + y$ will turn out to be a number, because of (T13) and (T14). Great!

A. Now II and III depend on each other, so we can combine them into a single statement like we did before.

B. Good point. Let's see, if I write $IV(x, y, z)$ to stand for the combined statement $II(x, y, z)$ and $III(x, y, z)$, my lists show that $IV(x, y, z)$ depends on

$$IV(y, X_L, z), \quad IV(x, y, Z_L), \quad IV(z, Z_L, y),$$
$$IV(Y_R, x, z), \quad IV(x, y, Z_R), \quad IV(Z_R, z, x),$$
$$IV(X_L, y, z), \quad IV(x, Y_R, z).$$

I think it was a good idea to introduce this new notation, like I(x, y) and so on, because it makes the patterns become clearer. Now all we have to do is find some way to rig up an induction hypothesis that goes from these six things to IV(x, y, z).

A. But uh-oh, it doesn't work. Look, IV(x, y, z) depends on IV(z, z_L, y), which depends on IV(y_R, y, z), which depends on IV(z, z_L, y) again; we're in a loop. It's the same stupid problem I noticed before, and now we know it's critical.

B. (pounding the dirt) *Oh no!* . . . Well, there's one more thing I'll try before giving up. Let's go all the way and prove a more general version of (T13):

> V. if $\quad x \le x' \quad$ and $\quad y \le y'$,
> then $\quad x + y \le x' + y'$.

This is what we really are using in our proofs, instead of doing two steps with (T13). And it's symmetrical; that might help.

A. We'll also need a converse, generalizing (T14).

B. I think what we need is

> VI. if $\quad x + y \ge x' + y' \quad$ and $\quad y \le y'$,
> then $\quad x \ge x'$.

A. Your notation, primes and all, looks very professional.

B. (concentrating) Thank you. Now the proof of V(x, x', y, y') depends on

> VI(X_L, x', y, y'),
> VI(Y_L, y', x, x'),
> VI(x, X'_R, y, y'),
> VI(y, Y'_R, x, x').

Hey, this is actually easier than the other one, the symmetry is helping.

Finally, to prove $VI(x, x', y, y')$, we need . . . the suspense is killing me, I can't think . . .

$$V(x, X'_L, y, y'), \qquad V(X_R, x', y, y').$$

A. (jumping up) Look, a *day-sum* argument, applied to the combination of V and VI, now finishes the induction!

B. (hugging her) We've won!

A. Bill, I can hardly believe it, but our proof of these two statements actually goes through for all *pseudo-numbers* x, x', y, and y'.

B. Alice, this has been a lot of work, but it's the most beautiful thing I ever saw.

A. Yes, we spent plenty of energy on what we both took for granted yesterday.

   I wonder if Conway himself had a simpler way to prove those laws? Maybe he did, but even so I like ours because it taught us a lot about proof techniques.

B. Today was going to be the day we studied multiplication.

A. We'd better not start it now, it might ruin our sleep again. Let's just spend the rest of the afternoon working out a proof that $-x$ is a number, whenever x is.

B. Good idea, that should be easy now. And I wonder if we can prove something about the way negation acts on pseudo-numbers?

14 THE UNIVERSE

B. (stretching) Good morning, love; did you think of any more mistakes in our math, during the night?

A. No, how about you?

B. You *know* I never look for mistakes. But a thought did strike me: Here we're supposed to have rules for creating all the numbers, but actually $\frac{1}{3}$ never appears. Remember, I was expecting to see it on the "fourth day," but the number turned out to be $\frac{1}{4}$. I kind of thought, well, $\frac{1}{3}$ is a little slow in arriving, but it will get here sooner or later. Just now it struck me that we've analyzed all the numbers, but $\frac{1}{3}$ still has never showed.

A. All the numbers that are created have a finite representation in the binary number system. I mean like $3\frac{5}{8} = 11.101$ in binary. And on the other hand, every number with a finite binary representation *does* get created, sooner or later. Like, $3\frac{5}{8}$ was created on the . . . eighth day.

B. Binary numbers are used on computers. Maybe Conway was creating a computerized world.

What's the binary representation of $\frac{1}{3}$ anyway?

A. I don't know, but it must have one.

B. Oh, I remember, you sort of do long division but with base 2 instead of 10. Let's see . . . I get

$$\tfrac{1}{3} = .0101010101 \ldots$$

and so on ad infinitum. It doesn't terminate, that's why it wasn't created.

A. "Ad infinitum." That reminds me of the last part of the inscription. What do you suppose the rock means about ℵ day and all that?

B. It sounds like some metaphysical or religious praise of the number system to me. Typical of ancient writings.

On the other hand, it's sort of strange that Conway was still around and talking, after infinitely many days. "Till the end of time," but time still hadn't ended.

A. You're in great voice today.

B. After infinitely many days, I guess Conway looked out over all those binary numbers he had created, and . . . Omigosh! I bet he *didn't* stop.

A. You're right! I never thought of it before, but the stone does seem to say he went right on. And . . . sure, he gets more numbers, too, because for the first time he can choose X_L and X_R to be infinite sets!

B. Perhaps time doesn't flow at a constant rate. I mean, to us the days seem like they're of equal length; but from Conway's point of view, as he peers into our universe, they might be going faster and faster in some absolute extra-celestial time scale. Like, the first earth day lasts one heavenly day, but the second earth day lasts only half a heavenly day, and the next is one fourth, and so on. Then, after a total of two heavenly days, zap! Infinitely many earth days have gone by, and we're ready to go on.

A. I never thought of that, but it makes sense. In a way, we're now exactly in Conway's position after infinitely many earth-days went by. Because we really *know* everything that transpired, up until ℵ day!

B. (gesticulating) *Another* plus for mathematics: Our finite minds can comprehend the infinite.

A. At least the countably infinite.

B. But the real numbers are uncountable, and we can even comprehend them.

A. I suppose so, since every real number is just an infinite decimal expansion.

B. Or binary expansion.

A. Hey! I know now what happened on ℵ day—the real numbers were all created!

B. (eyes popping) Migosh, I believe you're right.

A. Sure, we get $\frac{1}{3}$ by taking X_L to be, say,

$$\{.01, .0101, .010101, .01010101, \ldots\}$$

in binary notation, and X_R would be numbers that get closer and closer to $\frac{1}{3}$ from above, like

$$\{.1, .011, .01011, .0101011, .010101011, \ldots\}.$$

B. And a number like π gets created in roughly the same way. I don't know the binary representation of π, but let's say it's

$$\pi = 11.00100100001111 \ldots ;$$

we get Π_L by stopping at every "1,"

$$\Pi_L = \{11.001, 11.001001, 11.00100100001, \ldots\}$$

and Π_R by stopping at every "0" and adding 1,

$$\Pi_R = \{11.1, 11.01, 11.0011, 11.00101, \ldots\}.$$

A. There are lots of other sets that could be used for Π_L and Π_R, infinitely many in fact. But they all produce numbers equivalent to this one, because it is the first number created that is greater than Π_L and less than Π_R.

B. (hugging her again) So *that's* what the Conway Stone means when it says the universe was created on ℵ day: the real numbers are the universe.

Have you ever heard of the "big bang" theory the cosmologists talk about? This is it, ℵ day: Bang!

A. (not listening) Bill, there's *another* number also created on $\aleph$ day, a number that's not in the real number system. Take X_R to be empty, and

$$X_L = \{1, 2, 3, 4, 5, \ldots\}.$$

This number is larger than *all* the others.

B. Infinity! Outa sight!

A. I think I'll denote it by the Greek letter ω since I always liked that letter. Also $-\omega$ was created, I mean minus infinity.

B. $\aleph$ day was a busy, busy day.

A. Now the *next* day—

B. You mean $\aleph$ wasn't the end!

A. Oh no, why should Conway stop then? I have a hunch he was only barely getting started. The process never stops, because you can always take X_R empty and X_L to be the set of all numbers created so far.

B. But there isn't much else to *do* on the day after $\aleph$, since the real numbers fit together so densely. The noninfinite part of the universe is done now, since there's no room to put any more numbers in between two "adjacent" real numbers.

A. No, Bill; that's what *I* thought too, until you said it just now. I guess it just proves I like to argue with you. How about taking $X_L = \{0\}$ and $X_R = \{1, \frac{1}{2}, \frac{1}{3}, \frac{1}{4}, \frac{1}{5}, \ldots\}$. It's a number *greater* than zero and *less* than all positive real numbers! We might call it ϵ.

B. (fainting) Ulp . . . That's okay, I'm all right. But this is almost *too* much; I mean, there's gotta be a limit.

What surprises me most is that your number ϵ was actually created on $\aleph$ day, *not* the day after, because you could have taken $X_R = \{1, \frac{1}{2}, \frac{1}{4}, \frac{1}{8}, \frac{1}{16}, \ldots\}$. Also, there are lots of other crazy numbers in there, like

$$(\{1\}, \{1\frac{1}{2}, 1\frac{1}{4}, 1\frac{1}{8}, 1\frac{1}{16}, \ldots\})$$

which is just a hair bigger than 1.

And I suppose there's a number like this right next to all numbers, like π . . . no, that can't be . . .

A. The one just greater than π doesn't come until the day after $\aleph$. Only terminating binary numbers get an infinitely close neighbor on $\aleph$ day.

B. On the day after $\aleph$ we're also going to get a number *between* 0 and ϵ. And you think Conway was just getting started.

A. The neatest thing, Bill, is that we not only have the real numbers and infinity and all the in-betweens . . . we also have rules for telling which of two numbers is larger, and for *adding* and *subtracting* them.

B. That's *right*. We proved all these rules, thinking we already *knew* what we were proving—it was just a game, to derive all the old standard laws of arithmetic from Conway's few rules. But now we find that our proofs apply also to infinitely many unheard-of cases! The numbers are limited only by our imagination, and our consciousness is expanding, and . . .

A. You know, all this is sort of like a religious experience for me; I'm beginning to get a better appreciation of God. Like He's everywhere . . .

B. Even between the real numbers.

A. C'mon, I'm serious.

96

. . . .

B. I've been doing a few calculations with infinity. Like, rule
(4) tells us immediately that

$$\omega + 1 = (\{\omega, 2, 3, 4, 5, \ldots\}, \emptyset),$$

which simplifies to

$$\omega + 1 \equiv (\{\omega\}, \emptyset).$$

A. That was created on the day after $\aleph$ day.

B. Right, and

$$\omega + 2 \equiv (\{\omega + 1\}, \emptyset)$$

on the day after. Also,

$$\omega + \tfrac{1}{2} \equiv (\{\omega\}, \{\omega + 1\}).$$

A. What about $\omega - 1$?

B. $\omega - 1$! I never thought of subtracting from infinity, because a number less than infinity is supposed to be finite. But, let's grind it out by the rules and see what happens ... Look at that,

$$\omega - 1 \equiv (\{1, 2, 3, 4, \ldots\}, \{\omega\}).$$

Of course—it's the first number created which is larger than all integers, yet less than ω.

A. So *that's* what the Stone meant about an infinite number less than infinity.

Okay, I've got another one for you, what's $\omega + \pi$?

B. Easy:

$$\omega + \pi \equiv (\omega + \Pi_L, \omega + \Pi_R).$$

This was created on ... $(2\aleph)$ day! And so were $\omega + \epsilon$ and $\omega - \epsilon$.

A. Oho! Then there must also be a number 2ω. I mean, $\omega + \omega$.

B. Yup,

$$\omega + \omega = (\{\omega + 1, \omega + 2, \omega + 3, \omega + 4, \ldots\}, \emptyset).$$

I guess we can call this 2ω, even though we don't have multiplication yet, because we'll certainly prove later on that $(x + y)z \equiv xz + yz$. That means $2z \equiv (1 + 1)z \equiv z + z$.

A. Right, and

$$3\omega = (\{2\omega + 1, 2\omega + 2, 2\omega + 3, 2\omega + 4, \dots\}, \emptyset)$$

will be created on $(3\aleph)$ day, and so on.

B. We still don't know about multiplication, but I'm willing to bet that ω times ω will turn out to be

$$\omega^2 = (\{\omega, 2\omega, 3\omega, 4\omega, \dots\}, \emptyset).$$

A. Created on $\aleph^2$ day. Just imagine what Conway must be doing to all the smaller numbers during this time.

B. You know, Alice, this reminds me of a contest we used to have on our block when I was a kid. Every once in a while we'd start shouting about who knows the largest number. Pretty soon one of the kids found out from his dad that infinity was the largest number. But I went him one better by calling out "infinity plus one." Well, the next day we got up to infinity plus infinity, and soon it was infinity times infinity.

A. Then what happened?

B. Well, after reaching "infinityfinityfinityfinity . . ." repeated as long as possible without taking a breath, we sort of gave up the contest.

A. But there are still a lot more numbers left. Like

$$\omega^\omega = (\{\omega, \omega^2, \omega^3, \omega^4, \dots\}, \emptyset).$$

And still we're only at the beginning.

B. You mean, there's ω^ω, ω^{ω^ω}, and the limit of this, and so on. Why didn't I think of that when I was a kid?

A. It's a whole new vista . . . But I'm afraid our proofs aren't correct any more, Bill.

B. What? Not again. We already fixed them.
Oh-oh, I think I see what you're getting at. The day-sums.

A. Right. We can't argue by induction on the day-sums because they might be infinite.

B. Maybe our theorems don't even work for the infinite cases. It sure would be nice if they did, of course. I mean, what a feeling of power to be proving things about all these numbers we haven't even dreamed of yet.

A. We didn't have any apparent trouble with our trial calculations on infinite numbers. Let me think about this for awhile.

. . . .

It's okay, I think we're okay, we don't need "day-sums."

B. How do you manage it?

A. Well, remember how we first thought of induction in terms of "bad numbers." What we had to show was that if a theorem fails for x, say, then it also fails for some element x_L in X_L, and then it also fails for some x_{LL} in X_{LL}, and so on. But if every such sequence is eventually finite; I mean if eventually we must reach a case with $X_{LL...L}$ empty, then the theorem can't have failed for x in the first place.

B. (whistling) I see. For example, in our proof that $x + 0 = x$, we have $x + 0 = (X_L + 0, X_R + 0)$. We want to assume by induction that $x_L + 0$ has been proved equal to x_L for all x_L

in X_L. If this assumption is false, then $x_{LL} + 0$ hasn't been proved equal to x_{LL} for some x_{LL}; or, I guess, some x_{LR} might be the culprit. Any counterexample would imply an infinite sequence of counterexamples.

A. All we have to do now is show that there is no infinite ancestral sequence of numbers.

$$x_1, x_2, x_3, x_4, \ldots$$

such that x_{i+1} is in $X_{iL} \cup X_{iR}$.

B. That's a nice way to put it.

A. Also, it's true, because every number (in fact, every pseudo-number) is created out of *previously created* ones. Whenever we create a new number x, we could prove simultaneously that there is no infinite ancestral sequence starting with $x_1 = x$, because we have previously proved that there's no infinite sequence that proceeds from any of the possible choices of x_2 in X_L or X_R.

B. That's logical, and beautiful . . . But it almost sounds like you're proving the validity of induction, by using induction.

A. I suppose you're right. This must actually be an axiom of some sort, it formalizes the intuitive notion of "previously created" which we glossed over in rule (1). Yes, that's it, rule (1) will be on a rigorous footing if we formulate it in this way.

B. What you've said covers only the one-variable case. Our day-sum argument has been used for two, three, even four variables, where the induction for (x, y, z) relies on things like (y, z, x_L) and so on.

A. Exactly. But in every case, the induction went back to some *permutation* of the variables, with at least one of them getting an additional L or R subscript. Fortunately, this means that there can't be any infinite chain such as

$$(x, y, z) \rightarrow (y, z, x_L) \rightarrow (z_R, y, x_L) \rightarrow \cdots ,$$

and so on; if there were, at least one of the variables would have an infinite ancestral chain all by itself, contrary to rule (1).

B. (hugging her once again) Alice, I love you, in infinitely many ways.

A. (giggling) "How do I love thee? Let me count the ways." $1, \omega, \omega^2, \omega^\omega, \omega^{\omega^{\omega^{\cdots}}}, \ldots .$

B. It still seems that we have gotten around this infinite construction in a sneaky and possibly suspicious way. Although I can't see anything wrong with your argument, I'm still leery of it.

A. As I see it, the difference is between proof and calculation. There was no essential difference in the finite case, when we were just talking about numbers created before day $\aleph$. But now there is a definite distinction between proof and the ability to calculate. There are no infinite ancestral sequences, but they can be arbitrarily long, even when they start with the same number. For example, $\omega, n, n - 1, \ldots, 1, 0$ is a sequence of ancestors of ω, for all n.

B. Right. I've just been thinking about the ancestral sequences of ω^2. They're all finite, of course; but they can be so long, the finiteness isn't even obvious.

A. This unbounded finiteness means that we can make valid proofs, for example, that $2 \times \pi \equiv \pi + \pi$, but we can't necessarily calculate $\pi + \pi$ in a finite number of steps. Only God can finish the calculations, but we can finish the proofs.

B. Let's see, $\pi + \pi = (\pi + \Pi_L, \pi + \Pi_R)$, which ... Okay, I see, there are infinitely many branches of the calculations but they all terminate after finitely many steps.

A. The neat thing about the kind of induction we've been using is that we never have to prove the "initial case" separately. The way I learned induction, we always had to prove $P(1)$ first, or something like that. Somehow we've gotten around this.

B. You know, I think I understand the real meaning of induction for the first time. And I can hardly get over the fact that all our theory is really valid, for the infinite and infinitesimal numbers as well as the finite binary ones.

A. Except possibly (T8), which talks about the "first number created" with a certain property. We'd have to fix up a definition of what that means ... I suppose we could assign a number to each day, like say the largest number created on that day, and order the days that way ...

B. I sort of follow you. I've noticed that a number seems to be the largest created on its day when X_R is empty and X_L is all the previously created numbers.

A. Maybe that explains why there was $\aleph$ day and $(\aleph + 1)$ day, but no $(\aleph - 1)$ day.

B. Yeah, I guess, but this is all too deep for me. I'm ready to tackle multiplication now, aren't you?

16 MULTIPLICATION

A. Let's see that paper where you wrote down Conway's rule
   for multiplication. There must be a way to put it in symbols
   ... Hmm, we already know what he means by "part of the
   same kind."

B. Alice, this is too hard. Let's try to invent our own rule for
   multiplication instead of deciphering that message.

Why don't we just do like he did for addition. I mean, xy should lie between $X_L y \cup x Y_L$ and $X_R y \cup x Y_R$. At least, it ought to do this when negative numbers are excluded.

A. But that definition would be identical to addition, so the product would turn out to be the same as the sum.

B. Whoops, so it would . . . All right, I'm ready to appreciate Conway's solution, let's look at that paper.

A. Don't feel bad about it, you've got exactly the right attitude. Remember what we said about always trying things first?

B. Hah, I guess that's one lesson we've learned.

A. The best I can make out is that Conway chooses the left set of xy to be all numbers of the form

$$x_L y + x y_L - x_L y_L \qquad \text{or} \qquad x_R y + x y_R - x_R y_R,$$

and the right set contains all numbers of the form

$$x_L y + x y_R - x_L y_R \qquad \text{or} \qquad x_R y + x y_L - x_R y_L.$$

You see, the left set gets the "same kinds" and the right set gets the "opposite kinds" of parts. Does this definition make any sense?

B. Lemme see, it looks weird. Well, xy is supposed to be greater than its left part, so do we have

$$xy > x_L y + x y_L - x_L y_L?$$

This is like . . . yeah, it's equivalent to

$$(x - x_L)(y - y_L) > 0.$$

A. That's it, the product of positive numbers must be positive! The other three conditions for xy to lie between its left and right sets are essentially saying that

$$(x_R - x)(y_R - y) > 0,$$
$$(x - x_L)(y_R - y) > 0,$$
$$(x_R - x)(y - y_L) > 0.$$

Okay, the definition looks sensible, although we haven't proved anything.

B. Before we get carried away trying to prove the main laws about multiplication, I want to check out a few simple cases just to make sure. Let's see . . .

$$xy = yx; \tag{T20}$$
$$0y = 0; \tag{T21}$$
$$1y = y. \tag{T22}$$

Those were all easy.

A. Good, zero times infinity is zero. Another easy result is

$$-(xy) = (-x)y. \tag{T23}$$

B. Right on. Look, here's a fun one:

$$\tfrac{1}{2}x \equiv (\tfrac{1}{2}X_L \cup (x - \tfrac{1}{2}X_R), (x - \tfrac{1}{2}X_L) \cup \tfrac{1}{2}X_R). \tag{T24}$$

A. Hey, I've always wondered what *half of infinity* was.

B. Half of infinity! . . . Coming right up.

$$\tfrac{1}{2}\omega \equiv (\{1, 2, 3, 4, \ldots\},$$
$$\{\omega - 1, \omega - 2, \omega - 3, \omega - 4, \ldots\}).$$

It's interesting to prove that $\frac{1}{2}\omega + \frac{1}{2}\omega = \omega$ ... Wow, here's another neat one:

$$\epsilon\omega \equiv 1.$$

Our infinitesimal number turns out to be the reciprocal of infinity!

A. While you were working that out, I was looking at multiplication in general. It looks a little freaky for pseudo-numbers— I found a pseudo-number p for which $(\{1\}, \emptyset)p$ is not like $(\{0, 1\}, \emptyset)p$, even though $(\{1\}, \emptyset)$ and $(\{0, 1\}, \emptyset)$ are both equal to 2. In spite of this difficulty, I applied your Big Picture method and I think it is possible to prove

$$x(y + z) \equiv xy + xz, \qquad (\text{T25})$$
$$x(yz) \equiv (xy)z \qquad (\text{T26})$$

for arbitrary pseudo-numbers, and

$$\begin{aligned} &\text{if} \quad x > x' \quad \text{and} \quad y > y' \\ &\text{then} \quad (x - x')(y - y') > 0 \end{aligned} \qquad (\text{T27})$$

for arbitrary numbers. It will follow that xy is a number whenever x and y are.

B. Theorem (T27) can be used to show that

$$\text{if} \quad x \equiv y \quad \text{then} \quad xz \equiv yz \qquad (\text{T28})$$

for all numbers. So all of these calculations we've been making are perfectly rigorous.

I guess that takes care of everything it says on the tablet. Except the vague reference to "series, and quotients, and roots."

110

A. Hmm ... What about division? ... I bet if x is between 0 and 1, it'll be possible to prove that

$$1 - \frac{1}{1+x} \equiv$$
$$(\{x, \; x - x^2 + x^3, \; x - x^2 + x^3 - x^4 + x^5, \ldots\},$$
$$\{x - x^2, \; x - x^2 + x^3 - x^4, \ldots\}).$$

At least, this is how we got $\frac{1}{3}$, for $x = \frac{1}{2}$. Perhaps we'll be able to show that every nonzero number has a reciprocal, using some such method.

B. Alice! Feast your eyes on this!

$$\sqrt{\omega} \equiv (\{1, 2, 3, 4, \ldots\}, \left\{\frac{\omega}{1}, \frac{\omega}{2}, \frac{\omega}{3}, \frac{\omega}{4}, \ldots\right\}).$$

$$\sqrt{\epsilon} \equiv (\{\epsilon, 2\epsilon, 3\epsilon, 4\epsilon, \ldots\}, \left\{\frac{1}{1}, \frac{1}{2}, \frac{1}{3}, \frac{1}{4}, \ldots\right\}).$$

A. (falling into his arms) Bill! Every discovery leads to more, and more!

B. (glancing at the sunset) There are infinitely many things yet to do ... and only a finite amount of time ...

111

The reader may have guessed that this is not a true story. However, "J. H. W. H. Conway" does exist—he is Professor John Horton Conway of Cambridge University. The real Conway has established many remarkable results about these "extraordinal" numbers, besides what has been mentioned here. For example, every polynomial of odd degree, with arbitrary numbers as coefficients, has a root. Also, every pseudo-number p corresponds to a position in a two-person game between players Left and Right, where the four relations

$$p > 0, \quad p < 0,$$
$$p = 0, \quad p \parallel 0$$

correspond respectively to the four conditions

Left wins,	Right wins,
Second player wins,	First player wins,

starting at position p. The theory is still very much in its infancy, and the reader may wish to play with some of the many unexplored topics: What can be said about logarithms? continuity? multiplicative properties of pseudo-numbers? generalized diophantine equations? etc.

POSTSCRIPT

The late Hungarian mathematician Alfréd Rényi composed three stimulating "Dialogues on Mathematics," which were published by Holden-Day of San Francisco in 1967. His first dialogue, set in ancient Greece about 440 B.C., features Socrates and gives a beautiful description of the nature of mathematics. The second, which supposedly takes place in 212 B.C., contains Archimedes' equally beautiful discussion of the applications of mathematics. Rényi's third dialogue is about mathematics and science, and we hear Galileo speaking to us from about A.D. 1600.

I have prepared *Surreal Numbers* as a mathematical dialogue of the 1970's, emphasizing the nature of creative mathematical explorations. Of course, I wrote this mostly for fun, and I hope that it will transmit some pleasure to its readers, but I must also admit that I also had a serious purpose in the back of my mind. Namely, I wanted to provide some material which would help to overcome one of the most serious shortcomings in our present educational system, the lack of training for research work; there is comparatively little opportunity for students to experience how new mathematics is invented, until they reach graduate school.

I decided that creativity can't be taught using a textbook, but that an "anti-text" such as this novel might be useful. I therefore tried to write the exact opposite of Landau's *Grundlagen der Mathematik*; my aim was to show how mathematics can be "taken out of the classroom and into life," and to urge the reader to try his or her own hand at exploring abstract mathematical ideas.

The best way to communicate the techniques of mathematical research is probably to present a detailed case study. Conway's recent approach to numbers struck me as the perfect medium for illustrating the important aspects of mathematical explorations, because it is a rich theory that is almost self-contained, yet with close ties to both algebra and analysis, and because it is still largely unexplored.

In other words, my primary aim is not really to teach Conway's theory, it is to teach how one might go about developing such a theory. Therefore, as the two characters in this book gradually explore and build up Conway's number system, I have recorded their false starts and frustrations as well as their good ideas. I wanted to give a reasonably faithful portrayal of the important principles, techniques, joys, passions, and philosophy of mathematics, so I wrote the story as I was actually doing the research myself (using no outside sources except a vague memory of a lunchtime conversation I had had with John Conway almost a year earlier).

I have intended this book primarily for college mathematics students at about the sophomore or junior level. Within a traditional math curriculum it can probably be used best either (a) as supplementary reading material for an "Introduction to Abstract Mathematics" course or a "Mathematical Logic" course; or (b) as the principal text in an undergraduate seminar intended to develop the students' abilities for doing independent work.

Books which are used in classrooms usually are enhanced by exercises; so at the risk of destroying the purity of this "novel" approach, I have compiled a few suggestions for supplementary problems. When used with seminars, such exercises should preferably be brought up early in each class hour, for spontaneous class discussions, instead of being assigned as homework.

1. After Chapter 3. What is "abstraction," and what is "generalization"?

2. After Chapter 5. Assume that g is a function from numbers to numbers such that $x \leq y$ implies $g(x) \leq g(y)$. Define

   $$f(x) = (f(X_L) \cup \{g(x)\}, f(X_R)).$$

   Prove that $f(x) \leq f(y)$ if and only if $x \leq y$. Then in the special case that $g(x)$ is identically 0, evaluate $f(x)$ for as many numbers as you can. [*Note:* After Chapter 12, this exercise makes sense also when "numbers" are replaced by "pseudo-numbers."]

3. After Chapter 5. Let x,y be numbers whose left and right parts are "like" but not identical. Formally, let

   $$f_L: X_L \to Y_L, \qquad f_R: X_R \to Y_R,$$
   $$g_L: Y_L \to X_L, \qquad g_R: Y_R \to X_R$$

   be functions such that $f_L(x_L) \equiv x_L$, $f_R(x_R) \equiv x_R$, $g_L(y_L) \equiv y_L$, $g_R(y_R) \equiv y_R$. Prove that $x \equiv y$. [Alice and Bill did not realize that this lemma was important in some of their investigations, they assumed it without proof. The lemma holds also for pseudo-numbers.]

114

4. After Chapter 6. When we are developing the theory of Conway's numbers, from these few axioms, is it legitimate to be using the properties we already "know" about numbers, in the proofs? (For example, the use of subscripts like $i - 1$ and $j + 1$, and so on.) [*Warning:* This may lead to a discussion about metamathematics for which the instructor may have to be prepared.]

5. After Chapter 9. Find a complete formal proof of the general pattern after n days. [This makes an instructive exercise in design of notations. There are many possibilities, and the students should strive to find a notation that makes a rigorous proof most understandable, in that it matches Alice and Bill's intuitive informal proof.]

6. After Chapter 9. Is there a simple formula telling the day on which a given binary number is created?

7. After Chapter 10. Prove that $x \equiv y$ implies $-x \equiv -y$.

8. After Chapter 12. Establish the value of $x \oplus y$ for as many x, y as you can.

9. After Chapter 12. Change rules (1) and (2), replacing $\nleq$ by $<$ in all three places; and add a new rule:

$$x < y \qquad \text{if and only if} \qquad x \leq y \qquad \text{and} \qquad y \nleq x.$$

Now develop the theory of Conway's numbers from scratch, using these definitions. [This leads to a good review of the material in the first chapters; the arguments have to be changed in several places. The major hurdle is to prove $x \leq x$ for all numbers; there is a fairly short proof, not easy to discover, which I prefer not to reveal here. The students should be encouraged to discover that the new $<$ relation is not identical to Conway's, with respect to pseudo-numbers (although of course it is the same for all numbers). In the new case, $x \leq x$ does not always hold; and if

$$x = (\{\{\{0\}, \{0\})\}\}, \emptyset),$$

we have $x \equiv 0$ in Conway's system but $x \equiv 1$ in the new one! Conway's definition has nicer properties but the new relation is instructive.]

10. After Chapter 13. Show how to avoid Alice and Bill's circularity problem another way, by eliminating $III(z, Z_L, y)$ and $III(Z_R, z, x)$ from the requirements needed to prove $II(x, y, z)$. In other words, prove directly that we can't have $z + y \leq z_L + y$ for any z_L.

11. After Chapter 14. Determine the "immediate neighborhood" of each real number during the first few days after $\aleph$ day.

12. After Chapter 15. Construct the largest infinite numbers you can think of, and also the smallest positive infinitesimals.

13. After Chapter 15. Does it suffice to restrict X_L and X_R to countable sets? [This is difficult but it may lead to an interesting discussion. The instructor can prepare himself by boning up on ordinal numbers.]

14. Almost anywhere. What are the properties of the operation defined by

$$x \circ y = (X_L \cap Y_L, X_R \cup Y_R)?$$

[The class should discover that this is *not* $\min(x, y)$! Many other operations are interesting to explore, e.g., when $x \circ y$ is defined to be

$$(X_L \circ Y_L, X_R \cup Y_R)$$
$$\text{or } (X_L \circ y \cup x \circ Y_L, X_R \cup Y_R)$$

and so on.]

15. After Chapter 16. If X is the set of *all* numbers, show that $(X, \emptyset)$ is not equivalent to *any* number. [There are paradoxes in set theory unless care is taken. Strictly speaking, the class of all numbers isn't a set. Cf. the "set of all sets" paradoxes.]

16. After Chapter 16. Call x a *generalized integer* if

$$x \equiv (\{x - 1\}, \{x + 1\}).$$

Show that generalized integers are closed under addition, subtraction, and multiplication. They include the usual integers n, as well as numbers like $\omega \pm n$, $\frac{1}{2}\omega$, etc. [This exercise is due to Simon Norton.]

17. After Chapter 16. Call x a *real number* if $-n < x < n$ for some (nongeneralized) integer n, and if

$$x \equiv (\{x - 1, x - \tfrac{1}{2}, x - \tfrac{1}{4}, \ldots\}, \{x + 1, x + \tfrac{1}{2}, x + \tfrac{1}{4}, \ldots\}).$$

Prove that the real numbers are closed under addition, subtraction, and multiplication, and that they are isomorphic to real numbers defined in more traditional ways. [This exercise and those which follow were suggested by John Conway.]

18. After Chapter 16. Change rule (1), allowing (X_L, X_R) to be a number only when $X_L \not\geq X_R$ and the following condition is satisfied:

> X_L has a greatest element if and only if X_R has a least element.

Show that precisely the real numbers (no more, no less) are created in these circumstances.

19. After Chapter 16. Find a pseudo-number p such that $p + p \equiv (\{0\}, \{0\})$. [This is surprisingly difficult and it leads to interesting subproblems.]

20. After Chapter 15 or 16. The pseudo-number $(\{0\}, \{(\{0\}, \{0\})\})$ is > 0 and $< x$ for all positive numbers x. It's *really* infinitesimal! But $(\{0\}, \{(\{0\}, \{-1\})\})$ is smaller yet. And any pseudo-number $p > 0$ is $> (\{0\}, \{(\{0\}, \{-x\})\})$ for some suitably large number x.

21. After Chapter 16. For any number x define

$$\omega^x = (\{0\} \cup \{n\omega^{x_L} \mid x_L \in X_L, \; n = 1,2,3, \ldots\},$$

$$\left\{\frac{1}{2^n} \omega^{x_R} \mid x_R \in X_R, \; n = 1,2,3, \ldots\right\}).$$

Prove that $\omega^x \omega^y = \omega^{x + y}$.

22. After Chapter 16. Explore the properties of the symmetric pseudo-numbers S such that

$$(P_L, P_R) \in S \qquad \text{if and only if} \qquad P_L = P_R \subseteq S.$$

In other words, the elements of S have identical left and right sets, and so do the elements of their left and right parts. Show that S is closed under addition, subtraction, multiplication. Explore further properties of S (e.g., how many unlike elements of S are created on each day, and is their arithmetic interesting in any way?). [This open-ended problem is perhaps the best on this list, because there is an extremely rich theory lurking here.]

I will send hints to the solutions of exercises 9, 19, and 22 to any bona fide teachers who request them by writing to me at Stanford University.

Now I would like to close this postscript with some suggestions addressed specifically to teachers who will be leading a seminar based on this book. (All other people, please stop reading, and close the book at once.)

Dear Teacher: Many topics for class discussion are implicit in the story. The first few chapters will not take much time, but before long you may well be covering less than one chapter per class hour. It may be a good idea for everyone to skim the whole book very quickly at first, because the developments at the end are what really make the beginning interesting. One thing to stress continually is to ask the students to "distill off" the important general principles, the *modus operandi*, of the characters. Why do they approach the problem as they do, and what is good or bad about their approaches? How does Alice's "wisdom" differ from Bill's? (Their personalities are distinctly different.) Another ground rule for the students is that they should check over the mathematical details which are often only hinted at; this is the only way they can really learn what is going on in the book. Preferably they should tackle a problem first themselves before reading on. When an ellipsis, i.e., "..." appears, this often means the characters were thinking (or writing), and the reader should do the same.

When holding class discussions of such exercises as these, I have found it a good rule to limit the number of times each person is allowed to speak up. This keeps the loquacious people from hogging the floor and ruining the discussion; everybody gets to participate.

Another recommendation is that the course end with a three or four-week assignment, to write a term paper that explores some topic not explicitly worked out in the book. For example, the open-ended exercises in the above list illustrate several possible topics. Perhaps the students can do their research in groups of two. The students should also be told that they will be graded on their English expository style as well as on the mathematical content; say 50–50. They must be told that a math term paper should *not* read like a typical homework paper. The latter is generally a collection of facts in tabular form, without motivation, etc., and the grader is supposed to recognize it as a proof; the former is in prose style like in math textbooks. Another way to provide experience in writing is to have the students take turns preparing resumés of what transpires in class; then all the other students will be able to have a record of the discussions without being distracted by taking notes themselves. In my opinion the two weaknesses in our present mathematics education are the lack of training in creative thinking and the lack of practice in technical writing. I hope that the use of this little book can help make up for both of these deficiencies.

Stanford, California   D.E.K.
May 1974